Marc Achtelig

Basiswissen Technische Dokumentation
„How to Write That F*ing Manual"**

Ohne Umschweife zu benutzerfreundlichen Handbüchern und Hilfen

Erste Auflage
Zweisprachig: Englisch + Deutsch

indoition

indoition publishing e.K.
Goethestr. 24
90513 Zirndorf bei Nürnberg

Tel.: *+49 (0)911/60046-659*
Fax: *+49 (0)911/60046-863*
E-Mail: *info@indoition.de*
Internet: *www.indoition.de*

Autor: Marc Achtelig
Lektorat: Andrea R. Winter
Korrektorat: Elizabeth Meyer zu Heringdorf, Johannes Schubert
Satz: Marc Achtelig
Titeldesign: Marc Achtelig
Titelbild: E. Wodicka
Druck: Lightning Source

Marken
Alle in diesem Buch erwähnten, dem Verlag und Autor als Marken oder Warenzeichen
bekannten Begriffe, wurden in Textabschnitten in englischer Sprache durch entsprechende
Großschreibung gekennzeichnet. In Textabschnitten in deutscher Sprache erfolgte analog
ebenfalls Großschreibung. Weder Verlag noch Autor übernehmen jedoch eine Gewähr für
die Richtigkeit und Vollständigkeit dieser Kennzeichnung. Die Verwendung eines Begriffes
in diesem Buch lässt nicht auf das Bestehen oder Nichtbestehen eines markenrechtlichen
Schutzes schließen.

Warnhinweise und Haftungsausschluss
Die Informationen in diesem Buch wurden mit größtmöglicher Sorgfalt erstellt und auf
Richtigkeit und Vollständigkeit geprüft. Weder Verlag noch Autor übernehmen jedoch eine
Gewähr für Richtigkeit und Vollständigkeit gleich welcher Art. Ebenso übernehmen weder
Verlag noch Autor eine Gewähr für Gültigkeit und Anwendbarkeit der enthaltenen Inhalte.
Weder Verlag noch Autor übernehmen eine Haftung oder sonstige Verpflichtung gegenüber
natürlichen oder juristischen Personen im Hinblick auf Schäden oder Verluste jeder Art, die
als Folge aus den in diesem Buch gegeben Informationen entstehen, oder die aus dem
Fehlen von Informationen entstehen. Das Buch bietet keine individuelle Beratung,
insbesondere keine rechtliche Beratung. Stellen Sie vor Auslieferung Ihrer Produkte und vor
Veröffentlichung Ihrer Inhalte sicher, dass Sie alle Standards, Normen, Gesetze und
sonstigen Regelungen befolgen, die für Ihr eigenes Land sowie für alle Länder in die Sie Ihre
Produkte verkaufen und liefern maßgeblich sind. Alle in diesen Standards, Normen,
Gesetzen und sonstigen Regelungen enthaltenen Bestimmungen haben Vorrang gegenüber
den in diesem Buch gegebenen Empfehlungen.

Bibliografische Information der Deutschen Nationalbibliothek
Die Deutsche Nationalbibliothek verzeichnet diese Publikation in der deutschen
Nationalbibliografie; detaillierte bibliografische Daten sind im Internet über
http://dnb.d-nb.de abrufbar.

ISBN 978-3-943860-01-6

Über den Autor

Marc Achtelig ist Diplom-Ingenieur (FH) für Verfahrenstechnik und Wirtschaftsingenieurwesen und seit 1989 im Bereich der Technischen Kommunikation tätig.

Nach mehrjähriger Tätigkeit bei einem der größten deutschen Dokumentations-Dienstleister als Technischer Redakteur, „Information Architect" und Berater gründete er 2004 sein eigenes Consulting- und Dienstleistungsunternehmen.

Marc Achtelig war bereits in den 90er Jahren einer der Pioniere im Bereich Single-Source-Publishing, dem Ansatz, gedruckte Handbücher und Online-Hilfen aus einer gemeinsamen Textquelle zu erzeugen. Zu seinen Referenzen zählen zahlreiche Fachartikel, mehrere Bücher sowie diverse Vorträge, Tutorials und Workshops auf nationalen und internationalen Tagungen.

Für individuelle Beratungen und Schulungen erreichen Sie Marc Achtelig unter *ma@indoition.de*.

Inhalt / Contents

1   How to use this book

Welcome to this book, your brief companion to creating clear user assistance.

This book has been designed for those who understand that good user assistance matters, but who have no time to read thick textbooks on technical writing and technical communication. While this book can't cover every detail, it *can* cover those 80% that make the difference.

What you will find in this book

This book will help you to avoid the most serious of the typical mistakes that developers and marketing professionals make when documenting their own products—and this book will show you how to do it better.

Each topic begins on a new page, so skimming through the book is easy. You don't have to read everything from start to finish. All topics are independent of each other. You don't have to read any particular topic to understand another one.

What you won't find in this book

The book provides clear, brief rules and unambiguous recommendations. No boring theory, no rarely needed stuff for perfectionists, no musings, no highbrow grammar terms.

> **ⓘ Important:** The book can't provide individual advice; in particular, it can't provide any legal advice. Before shipping your products and before publishing any content, make sure that you also follow all relevant standards, laws, and other regulations that are applicable for both your own country and for all countries in which you sell your products. All rules that are given in these standards, laws, and other regulations take precedence over the recommendations given in this book.

Understanding the structure

The order of topics in the table of contents reflects the order of tasks that you typically need to perform:

- First, you need to analyze the requirements and can then develop the global information architecture.
  This step is covered by the first major section: *Structuring* 13.

- When you know your goal, you can then begin to design and set up the templates for your documents, starting with the global page and screen layout, going down to paragraph styles and character styles in detail.
  This step is covered by the second major section: *Designing* 53.

- When you've set up the information architecture and prepared the templates for your documents, you're now ready for the main task: Creating content.
  This step is covered by the third major section: *Writing* 95.

This book is about perfection—but it isn't perfect

When reading this book, you might notice that we sometimes don't manage to follow our own advice. You might find typos, grammar mistakes, and things that could have been said more clearly. Ouch—sorry!

Believe us: We've tried hard to make everything perfect. We've used some of the best spelling checkers, grammar checkers, and writing enhancement software. We had the text double-checked by human editors. Yet, there are still some mistakes. Actually, no book is perfect.

What can you learn from this for your own documents?

- Don't be frustrated if you find errors in your own documents some time after the documents have been published. This is embarrassing, but it's normal.

- Don't think that you can eliminate mistakes completely. You can only minimize their number. Do so, but spend your resources wisely. Don't forget to optimize the contents, too.

- If you provide some real value to your readers, they will tolerate more errors than if you provide little value.

We hope that this book will provide enough value so that you will forgive us for the errors that we have made.

Have a good time reading and writing.

2 Structuring

Draft a structure for your documents at an early stage. Don't start writing "intuitively" and let the documentation "evolve." If you start writing without having a plan for the structure, the whole process will take longer and will be more expensive.

When working on a plan, revising a decision costs nothing: All you have to do is to change your mind. Later, however, when you're working on the actual document, revising a decision will be costly. You will then have to change everything that you've produced so far. The more you've written, the more the changes cost in terms of time and money.

Structural decisions

All structural decisions have in common some *General structuring principles* 15. In addition to these general rules, setting up the structure involves the following specific decisions:

- **Media**
  Should you provide a printed user manual (PDF), or context-sensitive online help, or both? What information should go into the user manual, and what information should go into online help? Which help format should you use: plain HTML or some special help format? Can context-sensitive help calls be implemented? Should you provide interactive features and social features?

- **Documents**
  Should you put all information into one document, or should you supply several user manuals for specific purposes and user groups? How should you name your documents?

- **Document sections**
  What are the major sections that your documents should consist of? Are there any standard sections that you shouldn't forget?

- **Topics**
  What types of information do your clients need? How should you build and name the individual topics within the document? See *Which topics and headings?* 31

- **Order of sections and topics**
  How should you organize the sections and topics within your documents? See *Which order of topics?* 45

- **Navigation**
  Which navigational devices should you provide in printed documents and in online help systems? Where should you provide links or cross-references and where not?

2.1 General structuring principles

Good structure is even more important than good content. It doesn't matter how excellent your content is: If you have a poor structure, nobody will find and read this excellent content.

Universal principles

While some of the more specific rules apply only to particular structural levels such as documents, sections, or topics, the most important basic structuring rules apply universally:

- *Make it goal-centered* 16
- *Layer information* 22
- *Split complex information* 28

2.1.1 Make it goal-centered

It may sound trivial, yet it's the most frequently made mistake: When you write a user's guide, don't forget the user.

Set up a structure that reflects the users' *goals*. Don't set up a strictly systematic structure that reflects how your product works, or how the controls of your product are organized.

Don't tell users what *you* what to say. Tell users what *they* want to know.

Examples

The first example shows an excerpt from a user manual for a DVD recorder. Users don't want to know which connectors, controls, and other items are included in your product. Users want to know how to accomplish some specific task.

✖ **No:**

Contents

✔ **Yes:**

Contents

The next example shows an excerpt from a user manual for a program that manages client information.

✖ No:

Contents

File menu ....

   New ....

   Open ....

   Close ....

   Save ....

   Save as ....

   Export ....

   Print ....

   Exit ....

   ...

Edit menu

   Find ....

   Sort ....

   Filter ....

   ...

View menu ....

   ...

Tools menu ....

   Options ....

   Administration ....

   ...

✔ **Yes:**

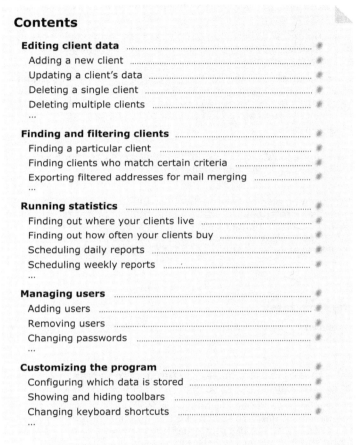

Contents

Why goal-based structure is important

A user-centered, goal-based structure is one of the key factors for successful user assistance and user-friendly documentation.

More than 90% of users don't care about how a product works. Most users don't read *any* documentation before they begin using the product.

- Users go to the documentation only when they feel stuck.
- Users stay in the documentation only until they feel unstuck.

Imagine the following, typical scenario:

A user is writing a letter and wants to add a picture. So the user's goal is to insert this picture into the letter.

- A goal-based document would provide a topic such as "Adding pictures to your documents," which is exactly what the user will be looking for when skimming the table of contents.

- A product-based document, on the contrary, would contain topics such as "The File Menu," "The Edit Menu," and "The Image Menu." The user must then guess where the needed function could be hidden. In the worst case, the user must open and read several irrelevant topics before finding the needed information.

Help people to *use* your product; don't describe the features of your product. Provide *assistance*—don't provide documentation.

Note:
A goal-based structure can also be an effective instrument to market your product. Guess what's more appealing when a potential client downloads a trial version of a program or a PDF manual before buying a product: a table of contents that lists all controls and menu items of the product, or a table of contents that reflects exactly what the potential client is aiming to do?

Goals may differ from tasks

Often, users' goals aren't identical with the tasks that need to be performed to achieve these goals.

For example:

- Your own goal is to enable users to use your product. Your task to achieve this goal is to write a manual (or some other form of user assistance).

- A user's goal may be to save paper when printing reports. However, the tasks that the user must perform to achieve this goal are different. For example, the user can change the printer settings so that two pages are printed onto one sheet of paper (task 1), or the user can enable double-sided printing (task 2).

When possible, use goals for headings rather than tasks because goals are what users are primarily looking for when browsing through your document. Users know what they want to achieve, but they *don't* know which tasks may be involved. Only use tasks for headings if users need to perform more than one task to achieve a goal.

Often, the best solution is to set up one topic that reflects the goal, and then to add subtopics that reflect the tasks necessary to achieve this goal.

Example:

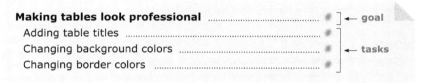

Where you shouldn't use a goal-centered structure

Goal-centered topic titles and structures are the optimal choice for procedural information, which usually accounts for most topics within a user manual or help system.

For conceptual information and for reference information, however, other structure models are often more appropriate (see also *Distinguish information types* 34 and *Primary structure models* 48). In this case, find topic titles that are in line with the chosen structure model.

2.1.2   Layer information

The users of a product don't read technical documentation for fun.

- Users want to find what they need to know as quickly as possible.
- Users want to stop reading as soon as possible and get back to work.

Don't bother your readers with any information that they don't need.

Layer your information according to your readers' needs.

Example

The following example illustrates a combination of some frequently used layering techniques:

- Each topic within the document belongs to a dedicated information type: "Concept," "Task," or "Reference."
- In the table of contents, each information type has a specific icon.
- The links to related information at the end of the topic are grouped according to the same information types.
- General information that's important for all users is always visible. It starts with the most important points.
- Additional information that's only important for a minority of users or in a minority of cases is put aside in expandable sections. If users want to see it, they must explicitly unhide it.
- Different aspects of information are layered in tabs.

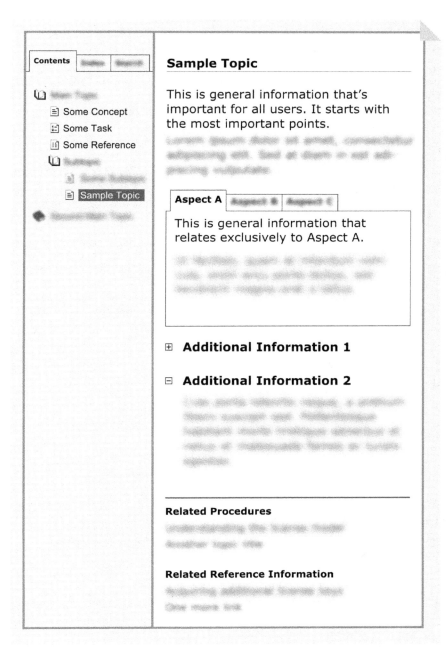

Contents

- Some Concept
- Some Task
- Some Reference

- Sample Topic

Sample Topic

This is general information that's important for all users. It starts with the most important points.

Aspect A

This is general information that relates exclusively to Aspect A.

⊞ **Additional Information 1**

⊟ **Additional Information 2**

Related Procedures

Related Reference Information

General principle

How and where you should layer your information depends on your product and on your audience. Only what the users of your product want to know at the same time should end up on the same layer.

For example, users rarely need conceptual information, procedural information, and reference information all at once.

- A user who needs to perform a particular task wants step-by-step instructions, but no verbose general information.

- A user who is already in the middle of a procedure only needs a detail about a particular parameter, but no step-by-step instructions.

- A beginner needs basic concepts but may be overwhelmed by too many details.

If you cram all information into one block instead of layering it, each user must also read a significant proportion of text that doesn't interest him or her at all. As a result, users feel that the document doesn't really meet their needs and get upset about the time they've wasted.

If you layer the information, however, users only read what actually interests them for their specific situation. They feel that your document exactly answers their questions and will come back for more.

Layering criteria

The most frequently used layering criteria are:

- importance (*must know*, *should know*, and *optional*)

- information type (*concepts*, *procedures*, *reference information*, and *examples*)

- user groups (*developer*, *administrator*, and *end user*)

- expertise (*beginner*, *intermediate user*, and *advanced user*)

Layering on the document level

If you provide more than one document, layering can begin even on the document level. For example:

- If you provide a user's guide and an administrator's guide, for example, this layers your content according to different user groups.

- If you provide a user's guide and a reference guide, this layers your content according to different information types.

Only include the information in each document that's in line with the document's information type.

Layering on the topic level

When layering on the topic level, you create topics that exclusively contain information of the same information type.

The most frequently used topic types are:

- *concept* (provides basics; answers "why?")
- *procedure* (provides tasks; answers "how?")
- *reference* (provides details; answers "what?")

Depending on your particular project, you can create any custom topic types that make sense. A topic type makes sense if you expect that a significant proportion of users will be interested exclusively in the information of this type without looking at other information.

For example, when documenting an ice cream machine, having only one "reference" topic type may not be enough. Instead, you can create the custom topic types "recipe reference" and "technical reference."

Tip:
In general, don't use more than 3 or 4 different information types. Users must be able to distinguish them clearly, which is rarely the case when using more than 3.

To organize layered topics within the table of contents, you have two options:

- You can create separate chapters where you collect all concepts, all procedures, and all reference information.

- You can list the topics of different topic types next to each other, typically in the following order: concept > procedure > reference information. In most cases, this option is more user-friendly because it keeps related information close together.

Note:
Use headings that make clear to which topic type each topic belongs (see *Find meaningful headings* 37).

Example:

Printing

The influence of paper quality on print quality .... ← concept
Printing a single picture .... ← procedure
Printing multiple pictures .... ← procedure
Printer settings .... ← reference

Layering within a topic

Within a topic, you can create subsections to separate different layers of information.

Start each subsection with a descriptive subtitle that clearly communicates what the subsection is about. Readers can then easily skim the text just by reading the subtitles.

In online help, you can create expandable sections (toggles) as a layer for initially hiding information of minor importance:

- Create the topic so that information that *all* users need is always visible.
- Put all information that only *some* users need into an expandable section.

Another option in online help is the use of tabs, with each tab showing a different information layer.

You can use tabs, for example:

- to separate concepts from procedures and from reference information within the same topic
- to separate different types of reference information
- to separate different kinds of examples

Layering within the text

Many writing rules for writing clear paragraphs and sentences are essentially also about layering. The common principle is always the same: Confront the reader with as little irrelevant information as possible.

The most important layering rules on the text level are:

- Always start with the main point. Start with what *most* users need to know, what users need to know *early*, and what's *not optional*.
- Don't mix different subjects within one paragraph. If there's a new subject, start a new paragraph.

- Don't mix different ideas or actions within one sentence. If there's a new idea or action, start a new sentence.

- Avoid all sorts of parentheses and nested sentences. If a parenthetic remark actually is important, create a full, separate sentence. If the parenthetic remark isn't important, omit it.

Layering navigation

In online help, you can also layer links to related topics.

For example, if you have the three topic types "concepts," "procedures," and "reference information," instead of providing one long list of related topics, you can provide three shorter lists:

- one list with links to related concepts
- one list with links to related procedures
- one list with links to related reference topics

2.1.3 Split complex information

From your own experience, you probably know that it sometimes first seems impossible to accomplish a task, but when you do it step by step, it suddenly becomes manageable.

Follow the same approach in your documents. Don't challenge users with a degree of complexity that they can't handle. Instead, provide the information step by step:

- Break down long, complex tasks into smaller, more manageable parts.
- Break down complex conceptual information and complex reference information into small chunks that are easy to understand and easy to remember.

Example

✖ **No:**

Servicing your car

1.
2.
3.
4.
5.
6.
7.
8.
9.
10.

...

77.
78.
79.

✔ **Yes:**

Servicing your car

Step 1: Checking the tires
1.
2.
3.

Step 2: Checking the wipers
1.
2.
3.
4.

Step 3: Checking the lights
1.
2.
3.

...

Step ⑩ Exchanging the air filter element
1.
2.
③

What can be split?

You can split complex information on all levels, from big to small:

- Split a long manual into several shorter manuals.
- Split a long section within a document into several shorter sections.
- Split a long topic into several shorter topics.
- Split a long subsection within a topic into several shorter subsections.

- Split a long paragraph into several shorter paragraphs.
- Split a long sentence into two or three shorter sentences.

Building and labeling chunks

- Create as many chunks of information as necessary, but as few chunks as possible.
  When building your chunks, balance the advantage that small chunks are easy to understand with the disadvantage that small chunks may lose their coherence.
- Build self-contained chunks.
  Don't build chunks of information that can't stand on their own. Form your chunks of information so that readers who skim the document can understand the information without having to scroll up or having to page back. In each chunk of information, cover everything that the heading or subheading of this chunk promises that you will cover. If that's impossible, change the heading or subheading.
- Label chunks.
  Add headings or subheadings that clearly communicate what's covered within a chunk of information. For example, if you break down a procedure into subprocedures, add a descriptive label (subheading) to each subprocedure.

Ordering and linking chunks

- Order chunks.
  Arrange the chunks of information just as you would have ordered the information if it wasn't chunked. If the chunks go into the same topic, place one chunk after the other. If each chunk becomes a separate topic, place one topic after the other.
- In printed manuals, don't link chunks.
  In a printed manual, don't explicitly refer or link from one chunk to another. The fact that they're standing closely together, one after the other, automatically provides the necessary coherence.
- In online help, link chunks if they span topics.
  In online help, add related topic links between chunks that end up in different topics.
  - If the chunks are procedures, add a related topic link only to the previous and next steps.
  - If the chunks provide concepts or reference information, add related topic links to *all* other chunks.

2.2 Which topics and headings?

Organizing the content into the right chunks of information and finding descriptive but concise headings for these chunks of information is one of the most critical parts of creating user-friendly content. It's also one of the most difficult parts.

The primary rule is: Base your decisions on the typical scenario that most users will go through when reading your material. Most users don't read manuals or online help from start to finish. Most users read user assistance only when they face a specific task or problem. They browse the table of contents, look for a keyword in the index, or use search to find a topic title that promises to deliver the solution to their problem.

For this reason, you need to:

- Organize information in a way so that users can read the document selectively. Each user should only have to read that short piece of information that's relevant to his or her particular situation.

- Build self-contained topics that cover everything that their headings promise to cover. Don't send readers on a wild-goose chase. Your goal should be that every reader needs to read only one topic to get his or her question answered.

- Label topics in such a way so that users can select the right topic without any trial and error. Make the first guess a success.

Key principles

The key principles of building and labeling your topics are:

- Build topics that fully cover what their headings promise that they will cover.
  See *Make topics self-contained* 32.

- Take into account that users may need different types of information. For example, users want *either* a quick step-by-step instruction *or* details, but they rarely want both.
  See *Distinguish information types* 34.

- Label each topic with a heading that clearly communicates what the topic covers, and what kind of information and what level of detail users can expect.
  See *Find meaningful headings* 37.

- When writing the topic heading, take care that the heading is concise and easy to read.
  See *Tips for writing headings* 40.

- Don't create meaningless topics that don't have a message on their own but just tell readers what comes "next."

2.2.1 Make topics self-contained

Self-contained topics are vital for online help, but they also make printed manuals more user-friendly.

Build topics that are complete, which means that they fully cover what their headings promise to cover.

Within the topic text, don't use any statements that relate to a given sequence of topics because you can't presume that users have read any "previous" topics or will read any "subsequent" topics. It's OK to link to other topics for related information but don't rely on the fact that readers follow the order of topics that your table of contents suggests.

✖ **No:** *As you've seen in the previous chapter, you must first create a report before you can print it.*

✔ **Yes:** *You must first create a report before you can print it (see "Creating Reports" on page 78).*

Self-contained topics have many advantages:

- Self-contained topics are essential if you want to create online help and a printed manual from the same text base, or if you want to produce documentation for different product versions from the same text base. You can then omit any topic in a specific version without having any negative side effects.

- Writing self-contained topics forces you, the author, to be concise and focused. Sometimes this can be challenging, but it greatly enhances the quality of your writing.

- Self-contained topics are a one-stop knowledge shop. Users find everything that they need to know in their specific situation and context in one place. There's no need to follow links. There's no need to go on a wild-goose chase and to collect information from multiple places.

What's the right topic size?

Don't attempt to adhere to a specific topic size limit. Try to keep every topic as short as possible, but as comprehensive as necessary to answer what the heading promises that the topic will answer.

If a topic seems to be getting too long:

- Consider splitting the topic if there are different pieces of information that users are likely to need independently. Find new headings that clearly describe the contents of both topics. If you create online help, add related topics links between both topics.

- Layer the information. Add subheadings. In online help, create expandable sections (toggles). See *Layer information* 22 for details.

How much scrolling is acceptable?

If readers are truly interested in the content, up and down scrolling is no problem. However:

- You must gain the readers' interest with the content that's visible without scrolling.
- You must keep the readers' interest throughout the whole text, or they'll stop scrolling and leave the topic without even having seen the rest.

Therefore, good general rules are:

- Start each topic with the main information.
- Make sure that the main information is visible without scrolling.
- If it's possible, use expandable sections (toggles) to layer the information. If readers can grasp the structure of the topic at one glance, the topic isn't too long.
- The more important a piece of information is, the closer it should be positioned to the upper left corner of the window.
- Don't add any extra wide tables or extra wide pictures that require horizontal scrolling. Vertical scrolling is acceptable, but horizontal scrolling is not.

2.2.2 Distinguish information types

Users don't read manuals and help texts for fun. They need specific information and want to get this information quickly:

- Beginners need basic information and want to understand how the product works in general, but they don't want to be bothered with any details for experts.

- Users who are already working with the product want simple step-by-step instructions that help them get their job done as quickly as possible. They *don't* want to be bothered with basic information for beginners. If they need details, they want to get just these details and not encounter a vast choice of information.

- Advanced users don't need step-by-step instructions anymore. They often just want to look up a specific parameter or setting.

Clearly distinguishing information types is your key to giving users just the information they need in a specific situation.

Make the used information types clearly distinguishable by using headings that are typical for each specific information type (see *Find meaningful headings* 37). As an option, in online help you can also use different icons in the table of contents, or you can use different colors or topic layouts.

There are three standard information types:

- "Concept"

- "Task"

- "Reference"

In addition to these standard types, you can also define your own information types depending on your particular product and document.

Tip:
If it's not obvious which information type to use, this usually indicates a structural problem. You're probably trying to stuff information into one topic that should be split up. Consider splitting the topic into several topics and finding better topic titles.

Standard information type "Concept"

"Concept" topics provide basic information and describe how things are related. They contain overviews, definitions, rules, and guidelines.

Even if most users mostly need step-by-step instructions (the "Task" information type), explaining the concepts is often equally important, especially with more complex products. Understanding the basic concepts:

- lets users understand and optimize their workflow
- encourages and helps users to explore the product on their own so that they can perform many tasks without needing to go to the documentation
- helps users find solutions for tasks that aren't described in the documentation
- helps users to prevent problems
- helps users to resolve problems

In addition, users who have understood the basic concepts can better remember all the information that's provided in the topics of other information types.

The "Task" information type

"Task" topics describe how to accomplish a specific job, or how to achieve a specific goal.

They list a series of steps that users can follow to produce the intended outcome.

The "Reference" information type

"Reference" topics usually contain detailed factual material—often in the form of tables and diagrams.

The topics aren't designed to be read completely but designed so that users can look up specific information selectively.

Custom information types

In addition to the standard information types, it sometimes makes sense to create some custom information types for specific information. In particular, you can often create specialized information types for different kinds of reference information.

Example: Imagine that you're documenting a medical instrument that measures some particular blood parameter. You have two user groups who use the device: lab assistants and doctors. So you could design two types of reference information: "measurement reference," which contains information on how to set up measurements, and "medical reference," which contains information on how to interpret the measurements' results. Doctors will need information of both information types, lab assistants will usually only need information of the type "measurement reference."

If defining your own information types makes sense, don't hesitate to do so.

However, don't create more types than necessary, and never create more types than your audience will be able to tell apart.

Mixing information types

Although the basic idea of information types is *not* to mix different classes of information, it can sometimes make sense to combine them in a controlled manner. There's no problem with this as long as you do it purposefully and with the users' needs in mind.

Example: Imagine that you're creating a user manual for an innovative product. Your audience hasn't used a product like this before. In this case, it could make sense to briefly describe the purpose of each procedure or to give some other background information before the steps. So you would have a mixture of the two information types "Task" and "Concept."

Likewise, sometimes it makes sense to provide details on specific parameters right within a step-by-step instruction, in particular if setting the parameters is always required. This is a mixture of "Task" and "Reference" information types.

2.2.3 Find meaningful headings

Don't let your readers guess what's in a topic. Create headings that clearly communicate:

- the contents of a topic
- the topic's information type (concept, task, or reference)
- if there are different user groups, or if the topic is relevant only for advanced users: the audience of the topic

It's a good idea to learn from marketing: Use a heading that attracts the attention of readers. However, don't promise anything that the topic cannot deliver.

If users feel that a topic answers just what they thought it would answer, you've created a positive user experience.

If users feel that a topic doesn't answer what they thought it would answer, they will conclude that the whole document is useless and stop reading. Many of them will never, ever come back. This is the worst thing that can happen!

Be aware of the fact that readers don't always see a heading in its hierarchical context. For example, when you browse the table of contents in a book and there's a main heading "Printing" and the subheadings "Overview" and "Reports," it's obvious that the Overview topic is about printing and that the Reports topic is about printing reports. However, when users find the same topics via a link or via search in online help, the heading "Overview" appears isolated and practically says nothing at all. Users don't have any clue then that it's an overview about printing. With the Reports topic, the problem is very similar: Is it about reports in general? About designing reports? About creating reports? About sending reports by email? Headings that are more meaningful are "Printing Overview" and "Printing Reports." It's perfectly OK to repeat the major heading "Printing" within the headings of the subordinate topics.

In addition, in online help, topic titles are typically listed as the results of full-text search. If you have multiple topics with an identical title such as "Overview," users have no clue which "Overview" to choose.

Here are some typical examples of popular but meaningless and often ambiguous headings that you should definitely avoid:

- Overview
- General Information
- Guidelines
- Basics

- Procedures
- Results
- Reference
- Additional Information
- Miscellaneous

Use headings that are more descriptive, such as "Configuration Overview" instead of "Overview," or "Safety Guidelines" instead of "Guidelines."

How to communicate the information type via the heading

By using typical grammatical constructions and typical phrases, you can often effectively indicate whether a topic contains basic conceptual information, step-by-step procedures (tasks), or detailed reference information.

Typical titles are:

- for concepts: *... Basics*; *How ...*; *Where ...*; *When ...*; *Why ...*
- for tasks: *...ing ...*
  (for example, *Printing Reports*)
- for reference information: *... Reference*
  (for example, *Parameter Reference*)

Note:
Many writers use headings that begin with "How to ..." for topics that describe tasks. Using the gerund (...ing) form, however, is usually shorter. Also, it automatically moves the important words to the beginning of the phrase. Users then don't have to read the complete heading to grasp its meaning. Another advantage with using the gerund form is that when multiple headings are listed below each other in the table of contents, it doesn't happen that all headings start with the same phrase "How to...". See also *Tips for writing headings* 40.

 (DE)

Typische Titel im Deutschen:

- Für Informationstyp Concept:
  - *Grundlagen zu ...*
  - *Wie ...*
  - *Wo ...*
  - *Wann ...*
  - *Warum ...*

- Für Informationstyp Task:
  - *... + Infinitiv* (z. B. *Berichte drucken*)
  - *So ...* (z. B. *So drucken Sie einen Bericht*)
- Für Informationstyp Reference:
  - *...referenz*
  - *... zum Nachschlagen*
  - *... im Detail*

It's OK to use a question as a title

If you can anticipate what questions users will have, you can often use these questions as topic titles. The advantage of this approach is that you automatically create topics that exactly match the users' need for information. This motivates users to actually read your document and works particularly well in documentation for consumer products.

✖ **No:**   *Exerting influence on fuel consumption*

✔ **Yes:**   *How can you save fuel?*

Using questions as titles is also a good method if you expect that users may have reservations about the points that you make. You can then phrase these reservations as questions and deal with them directly.

If you don't want to use questions explicitly, you can transform them into statements. Often, you can even include the answer or the key message of the topic this way.

✔ **Yes:**   *Is it OK to use a question as a title?*

✔ **Top:**   *It's OK to use a question as a title*

✔ **Top:**   *Why it's OK to use a question as a title*

2.2.4 Tips for writing headings

Write your headings in a way that makes it easy to skim the table of contents for relevant information.

Make headings parallel

In technical writing, identical structures aren't a sign of weak style but a key principle to enhance readability.

Parallel structures make the content more predictable. Readers don't have to process a new structure but can instead attend to the words alone.

Try to keep headings within a chapter, section, or other unit grammatically parallel—especially those on the same level.

✖ **No:** *Washing of trucks*
Washing cars
How can you wash a motorcycle?
How to wash a bicycle

✖ **No:** *Washing trucks*
Washing cars
Washing a motorcycle
Washing a bicycle

✖ **No:** *Washing trucks*
Cleaning cars
Giving a wash to motorcycles
Cleansing bicycles

✔ **Yes:** **Washing trucks**
Washing cars
Washing motorcycles
Washing bicycles

Move the distinctive information to the beginning of the heading

When possible, position the most characteristic information at the beginning of a heading. This makes it easy for readers to quickly skim a table of contents for relevant information.

✖ **No:** *Database Setup*
Database Reorganization
Database Compression

✖ **No:** *How to Set Up the Database*
How to Reorganize the Database
How to Compress the Database

✔ **Yes:** *Setting Up the Database*
Reorganizing the Database
Compressing the Database

▬ (DE)

Während es im Englischen nicht immer möglich und sinnvoll ist, einen Satz entsprechend umzustellen, ist dies im Deutschen oft wesentlich einfacher und effektiver.

✖ **Nein:** *Bügeln ohne Dampf*
Bügeln mit Dampf
Bügeln mit Dampfstoß

✔ **Ja:** *Ohne Dampf bügeln*
Mit Dampf bügeln
Mit Dampfstoß bügeln

Use verbs rather than nouns

When possible, use verbs rather than nouns. By using verbs, your headings become shorter, more concrete, and more practical.

✖ **No:** *Configuration of the Server*

✔ **Yes:** *Configuring the Server*

▬ (DE)

Im Deutschen bringt das Vermeiden von Substantivierungen zusätzlich den Vorteil, dass dadurch meist automatisch die inhaltsunterscheidenden Begriffe an den Anfang der Überschrift rücken. Mehrere Überschriften beginnen dann nicht mehr mit demselben Wort. Das erleichtert beim Lesen das schnelle Unterscheiden und Auffinden einer Information.

✖ **Nein:** *Konfigurieren des Servers*
Konfigurieren des Clients
Konfigurieren der Datenbank

✔ **Ja:** *Server konfigurieren*
Client konfigurieren
Datenbank konfigurieren

Use short, strong verbs instead of longer, verbose forms:

✖ **No:** *Conducting a Data Analysis*

✔ **Yes:** *Analyzing Data*

Make titles concise

Titles must be concise but also comprehensible. Don't use full sentences, but include whatever articles and prepositions are necessary to make the meaning of the title clear.

Avoid beginning a heading with a needless article.

✖ **No:**   *The Parameters Tab*

✔ **Yes:**   *Parameters Tab*

Titles that appear on a high hierarchy level within the table of contents should be particularly short because longer titles make it difficult to grasp a document's structure from the table of contents.

Titles that appear on a lower hierarchy level can be slightly more verbose because:

- Users view these titles less often.

- Topics on a low hierarchy level are often highly specialized topics that need more explanation than general topics at the top level.

If it helps to improve the readability of headings, it's OK to use colons, parentheses, or dashes.

Use singular nouns

Readers usually act on one thing at a time, so use singular nouns in your headings. However, when a singular noun doesn't make sense, using the plural is OK.

✖ **No:**   *Creating New Documents*

✔ **Yes:**   *Creating a New Document*

✖ **No:**   *Printing Reports*

✔ **Yes:**   *Printing a Report*

✔ **Yes:**   *Managing Folders*

Don't be vague

Avoid beginning headings with "Using ..." or "Working with ...." These headings are often vague and don't communicate the topic information type clearly. Instead, try to find a heading that clearly describes either a task *or* a concept to be discussed.

✖ **No:**   *Using the Spelling Checker*

✔ **Yes:** *How the Spelling Checker Works* (if you want to describe the concept)

or:

Checking Your Spelling (if you want to describe the task)

Attract the right readers

A good title doesn't attract as many readers as possible—it attracts the right readers.

There's nothing wrong with motivating readers to read a topic by pointing out the benefits. However, don't promise more than the topic can actually deliver.

✖ **No:** *Batch printing*

✔ **Yes:** *Saving time with batch printing*

43

2.3 Which order of topics?

Find a structure that doesn't reflect your own information needs and your own mental modal, but that reflects the mental model, the tasks, the goals, and the priorities of the users.

The order of sections and topics within a document determines:

- which information users perceive to be the most important, and which information users perceive to be the least important
- the context in which particular information is presented
- the succession in which users who read the document from start to finish pick up the information

Note:
Even online help has some sort of structure. In addition to the table of contents, many online help systems also provide a browse sequence so that users can follow the suggested order of topics by clicking a link to the "next" topic. Basically, in online help, readers are free to go anywhere, anytime, but the table of contents and the browse sequence recommend a certain sequence for those who would like the guidance.

Key criteria

When deciding on the hierarchical and linear structure of your document:

- Keep the structure as simple as possible (KISS principle = Keep It Simple and Stupid).
  See *Keep the structure flat* 46.
- Don't order your topics arbitrarily. Deliberately decide on one particular structure model, or on a combination of structure models, and then use these models consistently.
  See *Primary structure models* 48.

2.3.1 Keep the structure flat

Keep the structure of user manuals and help files as flat as possible. Avoid complex hierarchies that have sections, subsections, subsubsections, subsubsubsections, and so on. If you have a heading titled "4.3.5.6.7 What to avoid," something is seriously wrong.

As a rule of thumb, avoid having more than 3 levels. It can be challenging to find a structure that's both flat and clear, but it *is* possible in more than 90% of all documents.

If your audience has a low educational level, keep the structure especially flat.

Advantages of flat structures

- A flat structure is easier to comprehend.
- Emotionally, a flat structure is less daunting than a hierarchically deep structure. A flat structure encourages even inexperienced readers to explore your document.
- In a flat structure, the total number of topics is smaller. You need fewer topics for the sole purpose of building up the hierarchy.

Disadvantages of flat structures

- In a flat structure, there may be a large number of subtopics under one main topic. This can sometimes be confusing.
- Due to the larger number of topics on the same level, finding a particular topic can sometimes take longer in a flat structure. This may be irritating, particularly for advanced users who have already acquired sufficient knowledge to understand and remember a deeper structure.

Tip:
To make even a long list of topics clear, follow the key rules for writing user-friendly headings. In particular, position the main keyword at the beginning of each heading, and make headings short and parallel (see *Write meaningful headings* 37 and *Tips for writing headings* 40).

Tips for keeping a structure flat

- Include related information under one common heading.
- Don't think scientifically—think pragmatically. Even if there *is* some

hierarchy, this doesn't mean that your document must reflect it. Feel free to simplify.

- Always question whether it's actually important for the reader to understand the hierarchy of a particular relationship. Often, things are organized hierarchically within a product, knowing and understanding this hierarchy, however, is only important for developers but not for users. If the hierarchical organization isn't important for users, don't reproduce it in your document.

- Instead of creating several topics, consider layering the same information within one single topic. This reduces the total number of topics (see *Layer information* 22).

  - When writing online help, use expandable sections.

  - When writing a printed manual, use subheadings.

2.3.2 Primary structure models

The right structure:

- makes it easy to find information for the first time
- makes it easy to find information again after some time has passed
- presents the information in a didactical order, step by step
- helps to shape the mental model of how the product works

There's no universal structure model that fits all scenarios. Always base your decision on the specific goals and mental models of your audience.

Often, the best solution is a combination of two or more models. You can, for example, use one model for all level 1 headings, and then use different models for the subtopics within each section (level 2 headings).

Try to be as consistent and parallel as possible.

General rules

Regardless of which structure model you choose:

- Begin with the most important information, and end with the least important information.
- Put the information that all readers need before the information that only a few readers need.
- Put basic knowledge before expert knowledge.
- Put what's easy before what's more difficult. (This is especially important for tutorials and getting started sections, where you should motivate users by giving them a sense of achievement as soon as possible.)
- Put things that happen often before things that happen rarely.
- Put general information before more specific information.
- Put concepts before tasks, and put tasks before reference information.
- Put what's required before what's optional.
- Keep related information as close together as possible.

1-click access to frequently needed information

When possible, provide 1-click access to frequently needed information. Put that information on a high hierarchy level that's:

- important

- needed frequently
- needed by many readers

Example:

If you expect that many users will use keyboard shortcuts, don't create a hierarchy like "Appendix > Expert Features > Keyboard Shortcuts" but include the topic "Keyboard Shortcuts" on the top level.

User-based structure

A user-based structure can reflect:

- user groups
- the expertise of users
- use cases or scenarios
- stages of use
- users' goals
- users' tasks

Tip:
User-based structures make documents user-friendly because they directly take into account the users' needs. For this reason, when you can apply a user-based structure, do so. Apply user-based structures in particular to procedures.

Goal-based structure or task-based structure

Often, you can find major goals that comprise a number of tasks. Try to find verbs and then put them into a hierarchical structure.

Example:

- *creating*, *changing*, and *deleting* can be grouped into *managing*
- *displaying*, *printing*, and *exporting* can be grouped into *outputting*

Time-based structure

A time based structure can reflect:

- the chronological order of things that are done by the product
- the chronological order of processes that happen within the product
- the chronological order of steps that users must perform

Tip:
Use a time-based structure in particular for procedures and process descriptions.

Product-based structure

A product-based structure can reflect:

- components or modules of a product
- operator controls on a device
- the menu structure of a program

You can arrange items:

- from left to right
- from top to bottom
- from the outside to the inside
- from big to small

Many writers use a product-based structure because the product-based structure is already "given" by the product, so it's easy to set up without much consideration.

Many readers, however, dislike product-based structures because they don't reflect what readers are looking for within the documentation.

Tip:
Use a product-based structure for reference information, but don't use it for concepts, procedures, and examples.

Topical structure

Many subjects lend themselves to natural division. For example, the world can be divided into continents, continents can be divided into countries, countries into districts, and so on.

A topical structure can guide the reader to the correct topic only if the reader already knows and understands the basic principle behind the structure. So topical structures require at least some basic, prior knowledge of the subject.

Within the main chapters of a topical structure, you often need a second criterion to organize the subchapters. In the example of the world, for instance, you could sort the continents, countries, and districts either alphabetically, by size, or by number of inhabitants.

Alphabetical or numerical structure

Alphabetical and numerical structures:

- have no logical and no didactical approach whatsoever, so they're inappropriate for conceptual information

- enable readers to find specific information quickly; however, finding information only works if readers already know the subject well enough to be able to look for the right terms

- only make sense when there's a large number of items

Tip:
Use alphabetical or numerical structures for reference sections when there's no other, more obvious structure. Also use alphabetical or numerical structures when you expect that most users will use a document or document section primarily like an encyclopedia.

3 Designing

You never get a second chance to make a first impression.

Don't underestimate the importance of design. Good design cannot compensate for poor content, but it's one of the key factors of successful technical communication. An attractively designed communication product motivates users to read it, much like an attractively wrapped present motivates its recipient to open it.

Readers stay longer in an attractive-looking document than they stay in an unattractive-looking one. Readers who are motivated and stay in the document longer are more likely to understand the given information. Users who understand the given information are more likely to use your product successfully.

A clear design and well-thought-out templates also motivate and help you, the author, to produce clear, user-friendly content.

When to design

Many people, especially developers, argue that the design of a document should be the final step before you ship your document. Theoretically, this is true; practically, however, we don't recommend this approach for a number of reasons:

- If you haven't prepared appropriate styles for paragraphs, characters, links, and tables, you will have to go through all of your texts again just to apply the formatting. If you set up your templates beforehand, you can assign the appropriate styles right away.

- While writing, you don't see what you get. If you don't see the final result, this makes it more difficult to assume the perspective of the reader and to decide whether your text is easy to comprehend.

- Writing an unformatted text is simply less motivating than writing an attractive-looking one.

Tip:
Usually, the most efficient approach is to design the template beforehand without attempting to make the design pixel perfect at this stage. Instead, aim for about 80% of the quality standard that you want to achieve. Then, start writing and iteratively improve the templates where necessary. When writing is finished, review the design and make final adjustments.

What you need to know

You don't need to be a graphic artist to create a professional-looking, user-friendly template for a user manual or online help file.

- *Layout basics* 55
  Shows you the overall principles that you should follow when designing a template.

- *Choosing fonts and spacing* 83
  Shows you the basic typographic conventions that you should set up.

3.1 Layout basics

When setting up a template, two things are equally important:

- The template must be user-friendly. It must make finding information easy, and it must make reading easy.
- The template must be writer-friendly. It must be easy and efficient to use, and it must automate formatting as much as possible.

Don't trade in one for the other.

Things that make a template user-friendly

A user-friendly template:

- looks appealing, professional, and trustworthy
- makes it easy to skim a document for specific information
- makes it easy to decide what information is important and what information is less important
- makes reading easy and doesn't distract the reader

Things that make a template writer-friendly

A writer-friendly template:

- provides only a manageable number of styles
- uses a well-thought-out convention for style names so that writers can easily remember style names
- provides keyboard shortcuts to efficiently assign formats
- provides styles that minimize the need to manually tweak line breaks and page breaks
- provides styles that can be changed later without having to revise the whole document

> ℹ **Important:** Don't underestimate the importance of an efficient template. Typically, user assistance needs to be updated frequently—in particular, software user assistance. Each template inefficiency will multiply with the number of updates you make. If your document is translated into foreign languages, inefficiencies will also multiply with the number of translations.

Key principles

The key rules for designing user-friendly and writer-friendly templates are:

-
- *Use clear and simple design* 57
- *Visually support skimming* 59
- *Align texts and objects to a design grid* 60
- *Avoid amateurish formatting techniques* 62
- *Automate line breaks and page breaks* 66
- *Create styles semantically* 72
- *Create styles hierarchically* 79

3.1.1   Use clear and simple design

Less is more.

Keep the design as simple and plain as possible. Don't distract the reader.

Don't create a design that's an end in itself. Create a design that helps the reader to retrieve and process the given information.

Avoid variety

- Use only a few colors. Use a different color only if it has a particular purpose. Reserve bold colors for important things.
- Use only a few fonts and font sizes. Use a different font or font size only if it has a particular purpose. Reserve bold, large type for important things and for headings.
- Try to get along with as few styles as possible.
- Use the same styles throughout the whole document.
- Use the same styles also for all other documents that relate to the same product.
- In pictures, always use the same line width.
- Always use the same positions when arranging objects. Align all objects on a common design grid.

Avoid redundancy

Omit every letter, symbol, line, or other object that doesn't actually add any value.

- Objects that convey some important or helpful information *do* add value.
- Objects that help readers to find or process information *do* add value.
- Objects that are just there to fill some empty space *don't* add value.

For example:

- Don't use background images.
- Don't put your company logo on each page.
- Don't mention the author's name on each page.
- Don't put a copyright notice on each page.
- Don't put a revision number or release date on each page.

- Don't put the document title on each page.

  A running header or footer that shows the title of the current section, however, can be helpful because it keeps readers oriented.

Avoid clutter

Don't overload your pages with too much information.

Use white space purposefully to direct the readers' attention, to group things that belong together, and to set reading pauses.

It's no problem if an uncluttered layout makes your document a bit longer. If you provide valuable content, readers accept that they need to scroll or turn pages. However, they don't accept documents that overwhelm or confuse them.

3.1.2 Visually support skimming

Provide as much orientation as possible.

Provide clear visual means that help readers to skim your document for the information they need.

Most readers don't read manuals; they scan them quickly until they find an item of interest. But even then they don't start reading. First, they scan the item to decide whether it's worth investing in the time and effort to read the text. Only if they expect to gain some benefit from reading do they actually read.

This applies to printed manuals and online help alike.

Things that make skimming easy

The most important things that you can do to make your documents skimming-friendly are:

- Distinguish headings clearly from the text.
- Likewise, distinguish subheadings clearly from the text. If your authoring tool supports it, consider placing subheadings into a margin column.
- Clearly distinguish different heading levels.
- Choose a clear and simple design with only a few colors and fonts. You can then better highlight important key words and key phrases.
- Provide orientation so that readers can easily identify their current position within the document. In a printed manual, add running headers and footers. In online help, consider providing a breadcrumb trail.
- Create paragraph styles that visually communicate the information type of a paragraph's content. For example, create paragraph styles that clearly set apart warnings, notes, and tips.
- Provide separate paragraph styles for lists and procedures.
- Make page numbers large enough so that all readers can clearly read them. Page numbers aren't dispensable; they're a key navigation tool.

3.1.3 Align texts and objects to a design grid

Don't position text frames and other objects arbitrarily.

Align all objects to a common design grid so that they have common distances from the page or screen margins.

Typical applications of a design grid

Some typical examples of objects that should be aligned to a common design grid are:

- graphical elements and text elements on a title page
- headers and footers that are aligned with the text column
- logos, dates, version numbers, and page numbers
- texts within a picture

For the same reason, always align tables and pictures to the left of the text column instead of centering them. All elements then have a common starting line on the left side of the page, which looks much more consistent and less cluttered than a mixture of left-aligned and centered elements.

Example

In the following example, the lines indicate the common alignments on the different page types of a user manual:

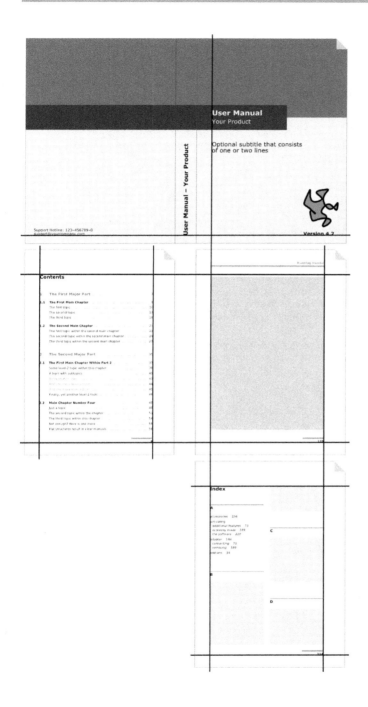

3.1.4   Avoid amateurish formatting techniques

Don't use empty paragraphs.

Don't use multiple space characters.

Don't use tabs.

Don't format your text manually. Only use the styles that are configured in your document template.

Problems with empty paragraphs

Don't use empty paragraphs to adjust the space above or the space below a paragraph. Use the appropriate paragraph settings instead.

Also, don't use empty paragraphs to create page breaks. If you later add or delete one or more lines from your text, everything will then move to a wrong position. Instead, always use proper manual page breaks; even better, when possible, completely automate page breaks (see *Automate line breaks and page breaks* 66).

All sorts of empty paragraphs seriously interfere with automatic page breaks. Look at the following example to see what may happen:

✖ No:

This is the preceding paragraph. Let's call it paragraph A.¶
¶
¶
¶

This is the heading

(page break)

¶
And here is the text that follows the heading. Initially, everything works well. The trouble starts when you later change the text. Imagine adding a few extra paragraphs above paragraph A so that paragraph A moves to the bottom of the page. As a result, the heading won't start properly at the top of the page.¶

This is the preceding paragraph. Let's call it paragraph A.¶

(page break)

¶
¶
¶

This is the heading

¶
And here is the text that follows the heading. Initially, everything works well. The trouble starts when you later change the text. Imagine adding a few extra paragraphs above paragraph A so that paragraph A moves to the bottom of the page. As a result, the heading won't start properly at the top of the page.¶

✔ Yes:

This is the preceding paragraph.¶

space above paragraph

This is a proper heading¶

space below paragraph

The heading has proper settings for the space above and for the space below the paragraph. So you don't have to add any empty paragraphs. If the heading moves to the top of a page, most authoring tools will remove the space above the paragraph automatically so that the heading starts correctly on top of the page.¶

This is the preceding paragraph.¶

(page break)

This is a proper heading¶

The heading has proper settings for the space above and for the space below the paragraph. So you don't have to add any empty paragraphs. If the heading moves to the top of a page, most authoring tools will remove the space above the paragraph automatically so that the heading starts correctly on top of the page.¶

Problems with multiple space characters

Don't use space characters to indent any text. Always configure proper indentation settings as part of the paragraph properties. If you use space characters for indenting text, as soon as you change a single word all the lines following the word may need to be edited manually.

In online documents that use a variable column width, indents with space characters don't work at all.

The following example shows how even a small change can break up your layout:

✖ No:

This·paragraph·isn't·indented.¶

···This·paragraph·is·indented·with·space·characters.¶

···This·paragraph·is·indented,·too.·But·it's·too·long. Now the trouble starts.¶

···You·could add·manual·line·breaks·and·even·more·space ←
···characters,·but·this·isn't·a·good·solution.·It·requires extra·work,←
···and·if·you·change·the·text,·everything may turn into a terrible ←
···mess.¶

···You·could·add·manual·line·breaks·and·even·more·space ←
···characters,·but·this·isn't·an·acceptable·solution.·It·requires·extra
·work,⊖
···and·everything·turns ⊖
···into·a·mess.¶

✔ Yes:

The·proper·way·is·to·define·a·paragraph·style·that has·a·left·indent.¶

left indent

The only time when using space characters to format your text makes sense is when you're quoting programming source code. When you copy and paste source code into your document, the source code is usually already indented with the help of space characters or tabs. Also, when formatting source code, you typically use a monospaced font, which has letters and characters that each occupy the same amount of horizontal space. This makes formatting with space characters easier.

Problems with tabs

Avoid using the tab key. Tabs may get you into trouble if you want to produce an online version of your document because tabs don't have any equivalent in HTML. Tabs then need to be converted into tables, space characters, or indents. This rarely works well.

Instead of using tabs, set up proper paragraph indentation, or use a borderless table. Even if you don't need to produce an online version of your document today, you may want to do so in the future.

Problems with customized styles

Don't customize paragraph settings and character settings for individual paragraphs. Always use the paragraph styles and character styles as they're defined in your document template.

If you overwrite a paragraph style for an individual paragraph or if you overwrite a character style for an individual word, this will get you into trouble when you need to change the design of your document later (which may seem unlikely today but often happens even if you don't anticipate it).

If you have exclusively used template styles and want to change the design, all you need to do is to change the style definitions. If you've applied manual formatting, however, you will need to update all manual formattings manually. If you have several documents—maybe even in different languages—this can result in many hours of extra work. So think ahead and apply formats with self-discipline.

3.1.5 Automate line breaks and page breaks

If you exclusively create online content, the only thing that you need to do is to use nonbreaking spaces where you want to prevent lines from breaking.

If you create printed manuals or PDF files, you also need to take care of hyphenation and page breaks. This can be very time-consuming if you don't automate your templates appropriately.

Be aware that every change that you make to your document may affect line breaks and page breaks. If you don't automate line breaks and page breaks the best that you possibly can, you'll have to revise them over and over again each time you update your document. Don't underestimate the frequency of document updates, especially in software user assistance.

If your line breaks and page breaks need manual tweaking, optimize them only as the final step after you have completely finished writing and editing.

Take a particularly close look at tables. Text within narrow table columns often needs some additional manual line breaks.

Where there should be line breaks

Follow these basic rules as closely as possible:

- In general, insert line breaks after punctuation marks instead of in the middle of a sentence.
- If a heading must have two lines, make the first line longer than the second line.
- Avoid line breaks that disrupt ideas that belong together.
- Don't insert a line break between the parts of a product name and between the parts of a company name.
- Don't insert a line break between a product name and the version number.
- Don't insert a line break between Internet addresses and email addresses.
- Don't insert a line break between numbers and units of measure.
- Avoid line breaks within function names, parameter names, commands, and quoted source code. Also turn off hyphenation here because readers might assume that the hyphen is part of the name.
- Avoid line breaks within cross-references.

✘ No:

Things that belong ↵
together shouldn't be ↵
separated by a line ↵
break.¶

✔ Yes:

Things that belong together ↵
shouldn't be separated ↵
by a line break.¶

Automating line breaks

To prevent unintentional line breaks, use all the automation options that your authoring tool provides.

- Add nonbreaking space characters instead of normal space characters (possible with most authoring tools).
- Define and use paragraph styles and character styles that don't allow hyphenation and line breaks (only possible in more advanced authoring tools). For example, create a special character style for your product name, and turn off hyphenation for this style.

Where there should be page breaks

Follow these basic rules as closely as possible:

- Try to keep related material on one page. If you can't avoid a page break, break pages so that readers can anticipate that an idea or topic continues on the next page.
- If you have a layout that has left pages and right pages, handle page breaks on right pages with special care. (Page breaks on left pages are less critical because readers can directly see the next page.)
- Don't separate headings from the text that follows.
- Avoid leaving major headings close to the bottom of the page.
- Avoid separating warnings, notes, and tips from the material that they concern.
- Never allow a page break within a warning.

- When possible, don't leave a single line of text at either the bottom of a page (a so-called "orphan") or on the top of a page (a so-called "widow"). If your authoring tool allows, set orphan control and widow control to a value of *2* for all paragraph styles.

- Don't leave a single list item or a single step of a procedure standing at the bottom of a page or at the top of a page.

- Don't separate an introductory phrase from the list, procedure, table, or figure that it introduces.

- Avoid breaking tables. If the table is too long, take care that at least two rows of the table are at the bottom and at the top of each page.

- If it isn't obvious what information each column of a table contains, repeat the column headings on the new page.

- If a table has a table title, keep the table title on the same page as the table. If you have a page break within the table, repeat the table title on the new page, followed the by the text "*(continued)*" in italic font style.

- If a figure has a figure title, keep the figure title on the same page as the figure.

- In an alphabetical index, if a main entry is followed by subentries, don't leave the main entry alone at the bottom of a column or page. If you must break up a list of subentries, repeat the main entry followed by the text "*(continued)*" in italic font style.

 (DE)

Im Deutschen ist für „*(continued)*" der Zusatz „*(Fortsetzung)*" üblich.

Automating page breaks

To automate page breaks, use all the advanced options that your authoring tool provides:

- Define paragraph style properties that automatically add a page break before the paragraph so that each paragraph of this style starts on a new page. You can use this for headings.

- Define paragraph style properties that always keep particular paragraphs together with the next paragraph on the same page. You can use this, for example, for phrases that introduce lists and procedures.

- Define paragraph style properties that prevent a page break from occurring within a paragraph. You can use this, for example, for warnings.

- Set up the number of widow lines and orphan lines that may be left at the top or bottom of a page, immediately before or after an automatic page break.

(DE)

Widow und *Orphan* haben im Deutschen keine wörtliche Entsprechung.

- Der analoge deutsche Begriff zu *Widow* (wörtlich „Witwe") lautet *Hurenkind* (letzte Zeile eines Absatzes erscheint als erste Zeile auf einer neuen Seite oder in einer neuen Spalte).

- Der analoge deutsche Begriff zu *Orphan* (wörtlich „Waisenkind") lautet *Schusterjunge* (erste Zeile eines Absatzes erscheint als letzte Zeile auf einer Seite).

Adding automatic page breaks before as many headings as possible can save you a lot of time.

- All chapters that are shorter than one page don't need any internal page breaks. So, there's nothing to worry about, nothing to check, and nothing to update.

- Changes within one chapter don't affect any of the following chapters because each of these chapters starts on a new page—just as before.

- Readers can spot chapters that begin on a new page more easily.

The downside of beginning a new page for each chapter is that it can make your document longer, so you need more paper. The shorter your average topic is, the stronger this effect will be.

For these reasons, good rules of thumb are:

- If your document is shipped electronically and you don't expect most readers to print it, start *all* headings on a new page.

- If your document is printed, start major headings on a new page (for example, levels 1 and 2), but start minor headings anywhere on a page. If most topics are shorter than half a page, only start heading level 1 on a new page.

Things that can't be fully automated

Usually, line breaks can be entirely automated. Page breaks, however, can only be automated to a degree of about 80% to 95%.

A need for a manual adjustment mostly results from a large image. When an image doesn't fit onto the rest of a page, the image is automatically shifted to the next page, leaving a lot of unused white space.

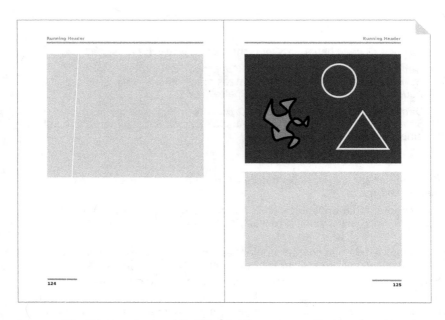

In cases like this, you have to decide whether you accept the unused, odd-looking white space or whether you resize the image to a smaller size so that it fits onto the previous page.

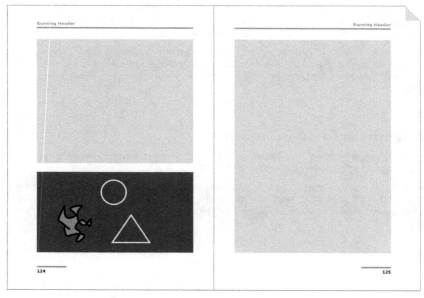

The advantage of resizing the image is that the visual appearance of your document improves.

Disadvantages of resizing the image are:

- Resizing needs extra work.

- The new image size may be inconsistent with the size of other images. For example, if you've used text in the image, the font size will be smaller than in other images.

- Your optimization may not be permanent. If you add, change, or delete some content before the image in the future, the image will move to a different position. You may then want to resize the image back to its original size. In addition, other images may now trigger poor page breaks, and the manual tweaking process starts all over again.

3.1.6   Create styles semantically

Use style names that describe the meaning of styles, not their visual appearance. For example, call a style "Emphasis" rather than "Arial 10 Point Bold." Semantic styles have some major advantages:

- The style name tells you when to use the style meaningfully.
- You can change the appearance of the style at any time without any negative side effects.
- Semantic styles are also a key prerequisite for structured authoring in XML and DITA. If you use semantic styles, your documents are well-prepared in case you want to shift to structured authoring someday.

To make your template writer-friendly, set up a well-thought-out naming convention for style names so that writers can remember them easily.

If your authoring tool supports it, define which paragraph style will be applied to the next paragraph when you press Enter.

Set up a keyboard shortcut for each style so that you can assign styles easily.

Tip:
If your authoring tool doesn't support keyboard shortcuts, use a third-party automation tool, such as the open source automation tool AutoHotkey.

Tips for naming conventions

- Use short style names. This makes it easy to spot a style in the style catalog of your authoring tool, on the status bar, and in other places.
- Create style names so that related styles are listed next to each other in the style catalog. You can achieve this by beginning the names of related styles with the same letters.

Tips for keyboard shortcut conventions

- Use keys that you can easily remember.
- Use the same scheme for all paragraph styles and another scheme for all character styles. For example, use [Alt]+[Shift]+letter for all paragraph styles, and use [Ctrl]+[Shift]+letter for all character styles.

Example scheme for paragraph styles

As a starting point, you can use or adapt the following scheme for your paragraph styles.

Note:
If your authoring tool doesn't allow you to define the suggested keyboard shortcuts, or if your authoring tool already uses these shortcuts, feel free to define other key combinations. The important thing is that you find a consistent scheme that you can easily remember.

Purpose	Suggested style names	Suggested keyboard shortcuts
Headings:		
heading level 1	h**1**	[Alt]+[Shift]+[**1**]
heading level 2	h**2**	[Alt]+[Shift]+[**2**]
heading level 3	h**3**	[Alt]+[Shift]+[**3**]
Subheadings:		
subheading that introduces a static section (always visible)	ss_**su**bhead_static	[Alt]+[Shift]+[**u**]
subheading that introduces an expandable section (toggle)	se_subhead_**e**xpandable	[Alt]+[Shift]+[**e**]
Body text:		
standard paragraph	bo_b**o**dy	[Alt]+[Shift]+[**o**]
Procedures:		
paragraph that introduces a procedure	pi_**p**rocedure_intro	[Alt]+[Shift]+[**p**]
step of a procedure (with number)	ps_procedure_**s**tep	[Alt]+[Shift]+[**s**]
indented first-level paragraph in a procedure (without number)	pp_procedure_plain	–
Lists:		
paragraph that introduces a list	li_**l**ist_intro	[Alt]+[Shift]+[**l**]
first-level item of a list (with bullet)	l1_list_level1_**b**ullet	[Alt]+[Shift]+[**b**]
indented first-level paragraph in	l1p_list_level1_plain	–

a list (without bullet)		
second-level item of a list (with bullet)	l2_list_level2_bullet	–
indented second-level paragraph in a list (without bullet)	l2p_list_level2_plain	–
Annotations: **Tips**	at_**t**ip	[Alt]+[Shift]+[**t**]
Annotations: **Notes**		
standard note	an_**n**ote	[Alt]+[Shift]+[**n**]
important note	ai_**i**mportant	[Alt]+[Shift]+[**i**]
Annotations: **Warnings**		
caution	ac_**c**aution	[Alt]+[Shift]+[**c**]
warning	aw_**w**arning	[Alt]+[Shift]+[**w**]
danger	ad_**d**anger	[Alt]+[Shift]+[**d**]

▬▬ (DE)

Wenn Sie lieber mit deutschen Bezeichnungen arbeiten, können Sie z. B. die folgenden Formatnamen und Tastenkürzel verwenden:

Zweck	Vorschlag für Formatnamen	Vorschlag für Tastenkürzel
Überschriften:		
Überschrift Ebene 1	ue**1**	[Alt]+[Shift]+[**1**]
Überschrift Ebene 2	ue**2**	[Alt]+[Shift]+[**2**]
Überschrift Ebene 3	ue**3**	[Alt]+[Shift]+[**3**]
Zwischenüberschriften:		
Zwischenüberschrift, die einen statischen Abschnitt einleitet (immer sichtbar)	zs_**z**ue_statisch	[Alt]+[Shift]+[**z**]
Zwischenüberschrift, die einen expandierbaren Abschnitt einleitet (Toggle)	ze_zue_**e**xpandierbar	[Alt]+[Shift]+[**e**]

Grundtext:		
Standardabsatz	tx_textkoerper	[Alt]+[Shift]+[x]
Handlungsanweisungen:		
Absatz, der eine schrittweise Handlungsanweisung einleitet	si_schritt_intro	[Alt]+[Shift]+[s]
Schritt einer schrittweisen Handlungsanweisung (mit Nummer)	sn_schritt_nummer	[Alt]+[Shift]+[n]
Eingerückter Fortsetzungsabsatz erster Ebene (ohne Nummer)	sf_schritt_folge	–
Listen:		
Absatz, der eine Liste einleitet	li_liste_intro	[Alt]+[Shift]+[l]
Listenpunkt erster Ebene einer Liste (mit Aufzählungszeichen)	l1_liste_ebene1_az	[Alt]+[Shift]+[a]
Eingerückter Folgeabsatz erster Ebene (ohne Aufzählungszeichen)	l1f_liste_ebene1_folge	–
Listenpunkt zweiter Ebene einer Liste (mit Aufzählungszeichen)	l2_liste_ebene2_az	–
Eingerückter Folgeabsatz zweiter Ebene (ohne Aufzählungszeichen)	l2f_liste_ebene2_folge	–
Anmerkungen: **Tipps**	at_tipp	[Alt]+[Shift]+[t]
Anmerkungen: **Hinweise**		
Standardhinweis	ah_hinweis	[Alt]+[Shift]+[h]
Wichtiger Hinweis	aw_wichtig	[Alt]+[Shift]+[w]
Anmerkungen: **Warnhinweise**		
Vorsicht	av_vorsicht	[Alt]+[Shift]+[v]
Warnung	aw_warnung	[Alt]+[Shift]+[w]
Gefahr	ag_gefahr	[Alt]+[Shift]+[g]

Example scheme for character styles

As a starting point, you can use or adapt the following scheme for your character styles.

Note:
If your authoring tool doesn't allow you to define the suggested keyboard shortcuts, or if your authoring tool already uses these shortcuts, feel free to define other key combinations. The important thing is that you find a consistent scheme that you can easily remember.

Purpose	Suggested style name	Suggested keyboard shortcut
To highlight the name of the documented product and to disable automatic hyphenation for the product name (optional).	cp_product	[Ctrl]+[Shift]+[p]
To highlight all user interface and interaction elements, such as window titles, menu items, buttons, keys, levers, and so on.	ce_element	[Ctrl]+[Shift]+[e]
To mark parameters, such as parameters of function calls, formulas, and so on.	ca_parameter	[Ctrl]+[Shift]+[a]
To mark product-specific technical terms.	ct_term	[Ctrl]+[Shift]+[t]
To emphasize words and expressions if this is necessary to avoid confusion.	cm_emphasis	[Ctrl]+[Shift] +[m]
To highlight important keywords that help readers to skim the text (optional).	cs_strong	[Ctrl]+[Shift]+[s]
To mark input that users must type.	ci_input	[Ctrl]+[Shift]+[i]
To mark quotes of source code and configuration files.	cc_code	[Ctrl]+[Shift]+[c]
To mark links / cross-references to other topics within the	cl_link	[Ctrl]+[Shift]+[l]

document as well as links to
external web sites and
documents.

━━ (DE)

Wenn Sie lieber mit deutschen Bezeichnungen arbeiten, können Sie z. B. die
folgenden Formatnamen und Tastenkürzel verwenden:

Zweck	Vorschlag für Formatnamen	Vorschlag für Tastenkürzel
Um den Namen des beschriebenen Produkts hervorzuheben sowie um die automatische Silbentrennung für den Produktnamen zu deaktivieren (optional)	zp_produkt	[Ctrl]+[Shift]+[p]
Um alle Schnittstellen- und Bedienelemente zu kennzeichnen, wie Fenstertitel, Menüpunkte, Schaltflächen, Knöpfe, Bedienhebel, usw.	ze_element	[Ctrl]+[Shift]+[e]
Um Parameter zu kennzeichnen, wie z. B. Parameter für Funktionsaufrufe, Formeln, usw.	za_parameter	[Ctrl]+[Shift]+[a]
Um produktspezifische Fachbegriffe zu kennzeichnen	zi_begriff	[Ctrl]+[Shift]+[i]
Um Wörter und Ausdrücke hervorzuheben, wenn diese Hervorhebung Missverständnissen vorbeugen kann	zb_betonung	[Ctrl]+[Shift]+[b]
Um wichtige Schlüsselwörter hervorzuheben, die den Lesern helfen, den Text schnell zu überfliegen (optional)	zh_hervorhebung	[Ctrl]+[Shift]+[h]
Zur Kennzeichnung von Benutzereingaben über die Tastatur	zg_eingabe	[Ctrl]+[Shift]+[g]
Zur Kennzeichnung von Quellcode-Zitaten und Zitaten	zc_code	[Ctrl]+[Shift]+[c]

aus Konfigurationsdateien		
Zur Kennzeichnung von Links / Querverweisen zu anderen Themen innerhalb desselben Dokuments, zu externen Webseiten oder externen Dokumenten	zl_link	[Ctrl]+[Shift]+[l]

3.1.7 Create styles hierarchically

If your authoring tool supports it, don't set up each style independently, but create a hierarchy of styles.

When styles are organized hierarchically, you can later change settings centrally in one place. All child styles inherit the changed setting automatically. For example, if you want to change the font, you only need to change the font of the root element. All other styles that don't use another special font then automatically inherit the new font as well. Thus, instead of having to adapt *all* styles manually, you only need to change *one* style.

Example of hierarchically organized styles

The basic styles (see *Create styles semantically* 72) could be organized as outlined below. If you want to change, for example, the color of all headings, you only need to change the color of the parent style **Headings**. You don't have to change the styles **h1**, **h2**, and **h3** individually.

Custom_Paragraph_Styles

- **Headings**
  - h1
  - h2
  - h3
- **Subheadings**
  - ss_subhead_static
  - se_subhead_expandable
- b0_body
- **Procedures**
  - pi_procedure_intro
  - ps_procedure_step
  - pp_procedure_plain
- **Lists**
  - li_list_intro
  - l1_list_level1_bullet
  - l1p_list_level1_plain
  - l2_list_level2_bullet
  - l2p_list_level2_plain
- **Annotations**
  - at_tip
  - **Notes**
    - an_note
    - ai_important
  - **Warnings**
    - ac_caution
    - aw_warning
    - ad_danger

Custom_Character_Styles

- cp_product
- ce_element
- ca_parameter
- ct_term
- cm_emphasis
- cs_strong
- ci_input
- cc_code
- cl_link

(DE)

Für die deutschen Formatnamen gilt entsprechend:

Eigene_Absatzformate

- **Überschriften**
  - ue1
  - ue2
  - ue3
- **Zwischenüberschriften**
  - zs_zue_statisch
  - ze_zue_expandierbar
- b0_textkoerper
- **Handlungsanweisungen**
  - si_schritt_intro
  - sn_schritt_nummer
  - sf_schritt_folge
- **Listen**
  - li_liste_intro
  - l1_liste_ebene1_az
  - l1f_liste_ebene1_folge
  - l2_liste_ebene2_az
  - l2f_liste_ebene2_folge
- **Anmerkungen**
  - at_tipp
  - **Hinweise**
    - ah_hinweis
    - aw_wichtig
  - **Warnhinweise**
    - av_vorsicht
    - aw_warnung
    - ag_gefahr

Eigene_Zeichenformate

- zp_produkt
- ze_element
- za_parameter
- zi_begriff
- zb_betonung
- zh_hervorhebung
- zg_eingabe
- zc_code
- zl_link

3.2 Choosing fonts and spacing

The chosen fonts and the font settings play a key role in the image that your document conveys.

Fonts and front sizes also have a major effect on readability.

Tip:
If you're unsure about a specific setting for a printed manual, don't hesitate to print a test page on your office printer. On paper, the visual appearance and readability are often very different than they are on screen. Put several alternatives next to each other and compare them. When in doubt, ask a second person. This person doesn't necessarily have to belong to your audience. Questions of readability are quite universal.

Key questions

Choosing fonts and setting font attributes involves the following questions:

- *Which font?* 84
- *Which font style?* 89
- *Which font size and font spacing?* 91
- *What line length?* 93

3.2.1 Which font?

If your company has design guidelines that require you to use a particular font, do so. If you ship your documents electronically, make sure that all users have the required fonts. If you have the proper license, embed the fonts, or ship them along with your product.

If your company doesn't have any corporate fonts, use a common font that's easily readable. Don't choose any stylish, trendy fonts. If you ship your documents electronically, use a font that's installed on the computers of all users. Bear in mind that the default fonts aren't identical on all operating systems.

Use as few fonts as possible (see *Use clear and simple design* 57).

Types of fonts

There are three main types of fonts:

- serif fonts
- sans-serif fonts
- monospaced fonts

This is a **serif font**.
The small embellishments at the
end of the letters are called "serifs."

This is a **sans-serif font**.
"Sans" is French and means "without."
So, this is a font without serifs.

This is a **monospaced font**. Fonts like this were
common on mechanical typewriters.
All letters have the same width. For example, the
"i" is just as wide as the "m." As a result, all
characters are precisely positioned below each other.
The space character also has the same width as all
other characters.

(DE)

Deutsche Begriffe: serifenlose Schrift, Serifenschrift, nichtproportionale Schrift

General recommendations

In general:

- Sans-serif fonts tend to look quite modern and innovative, whereas serif fonts look more traditional. Using one or the other can emphasize the image of your product into an intended direction.
- Sans-serif fonts usually have better readability—particularly on screen. For this reason, always use sans-serif fonts in online help.
- The serifs of serif fonts guide the eye along the line. This makes reading long texts slightly less tiring and is the reason why most novels have serif fonts. In user assistance, however, users usually read only short passages, so this advantage is rather unimportant here.
- Don't use monospaced fonts except for source code and user input.
- Avoid all sorts of squiggly fonts.
- Avoid fonts that have very thin lines. Many people have trouble reading them. On screen, thin lines may disappear completely.
- Avoid fonts that have very thick lines. They make your whole text look bold and obtrusive, and you lose the possibility of making individual words bold. In addition, fonts with thick lines need a lot of space, so you can put less information onto one page.
- Avoid fonts that have lowercase letters with a very small height. This is particularly important in online help, where the height of lowercase letters should be at least 50% of the height of capital letters to be clearly readable.
- If you ship your document electronically, make sure that the font is available on the users' computers, or make sure that you can embed the font into the document.
- Make sure that the fonts that you choose support all language-specific characters of the languages into which your document might be translated. Plan ahead. As business evolves, languages that may seem unlikely to be used today may soon become mission-critical.

Tip:
When in doubt, use the more common font. Most users find reading a familiar font more relaxing than reading an unfamiliar one.

Large and small fonts

In places where space is tight, you sometimes need to add very small text. However, when scaled down, the readability of many standard fonts is very poor. If your standard font looks blurred in cases like this, try a font that has taller lowercase letters. When possible, look for a font that's been particularly designed to be used in small sizes.

Likewise, many fonts aren't appropriate for large headings either. If a font looks inept here, try a narrower font or a font that has a smaller line width.

Combining fonts

In general, use as few fonts as possible.

Use an additional font only if this improves the readability of your document. This typically is the case if the additional font marks a special element, such as headings, labels, or computer source code. Also, you sometimes need a special font for extremely small text or for extremely large text.

The key rules for combining fonts are:

- Avoid mixing more than two fonts. Note that if you have a company logo or product logo on a page, there may be yet another font in the logo.

- Don't mix fonts that look very similar. Some readers won't notice; others might think that it's a mistake.

  However, do use fonts that have *some* characteristics in common so that the combination looks intentional. For example, combine two fonts that have equal line widths, two fonts that are both modern, two fonts that are both elegant, and so on.

- Use fonts that have identically sized lowercase letters.

- Finding a good combination of two serif fonts is especially difficult. Mixing two sans-serif fonts tends to be easier.

- A good option can be to mix a serif font with a sans-serif font. Often, for example, a serif font is used for body text, and a sans-serif font is used for headings. For some fonts, there's even both a serif version and a sans-serif version available, which blend especially well.

Bold and italic font styles

The standard function in many editors that makes text bold or italic does so by applying a mathematical algorithm to the used typeface. This usually produces acceptable results.

However, if you want to achieve a truly professional layout in a printed manual, don't apply the automatic functions; instead, assign a special bold version or italic version of the font that you use. For most professional fonts, special bold and italic versions are available, but they must often be purchased separately.

ⓘ **Important:** For online help, always use the standard bold and italic function. If you use a special font version, this version might not be available on the users' computers.

Recommended fonts

Avoid using Arial in body text because this font is very narrow and thus often results in poor readability and in an amateurish look. However, the fact that Arial is very narrow makes it an excellent choice for headings.

Some good, easily available fonts for the body text of printed manuals are:

- **Bitstream Vera** and **Bitstream Vera Sans** (initially developed for Linux; quite similar to Verdana)
- **Calibri** (sans-serif; ships with Windows)
- **Corbel** (sans-serif; ships with Windows)
- **Consolas** (monospaced; a good alternative to Courier; ships with Windows)
- **Constantia** (serif; designed to be used both on paper and on screen; ships with Windows)
- **Meta** (must be licensed separately)
- **Univers** (must be licensed separately)
- **Syntax** (must be licensed separately)

Common fonts for online help that are installed on most computers are:

- **Arial** (ships with Windows)
- **Verdana** (ships with Windows)
- **Segoe UI** (ships with Windows)
- **Trebuchet MS** (ships with Windows)
- **Courier New** (monospaced; ships with Windows)

In online help, generally use Verdana. Verdana has an excellent readability on screen and is installed on most computers. Use Arial only if you have little space—for example, in headings and in narrow table cells. Use Courier New for program source code.

In HTML, use the font-family attribute to specify some replacement fonts. If a particular font isn't available on a user's computer, the browser then uses the replacement font instead of the original font. (If you don't specify a replacement font, the browser uses its default font, which may be very different from your original font.)

If your authoring tool doesn't support adding the font-family attribute, consider running a global search and replace on the final HTML documents.

For Verdana, the font-family attribute may look like this:
```
font-family: 'Verdana', 'Bitstream Vera Sans', 'Trebuchet MS',
'Geneva', 'Arial', sans-serif;
```

Examples

Bitstream Vera	How do you like this?
Calibri	How do you like this?
Corbel	How do you like this?
Consolas	How do you like this?
Constantia	How do you like this?
Meta	How do you like this?
Univers	How do you like this?
Syntax	How do you like this?
Arial	How do you like this?
Verdana	How do you like this?
Segoe UI	How do you like this?
Trebuchet MS	How do you like this?
Courier New	How do you like this?

3.2.2   Which font style?

Avoid variety (see *Use clear and simple design* 57).

The fewer font styles you need, the better.

Use only one font attribute at a time. For example, make text bold, *or* italic, *or* underlined, but don't make it bold + italic + underlined.

Font style Bold

Bold text should *identify* information.

Make headings bold.

Within the body text, only make bold what you want to use as a visual label to support skimming.

Font styles Italic and Small Caps

It's harder to read italic text than to read normal text, especially when reading on screen.

Avoid long text in italics; use italics only to *emphasize* individual words or phrases.

Don't use small capitals at all. Their readability is very poor *both* in printed manuals *and* in online help.

 (DE)

Deutscher Begriff: Kapitälchen

Font style Underlined

In online help, use underlined text exclusively for hyperlinks. Even if links are not underlined in your document, don't use underlined text for anything else; many users will click the underlined text because they think that it's a link.

If you've underlined text in a printed manual, only use it for single words or short phrases. Long sections of underlined text are very difficult to read. What's more, if you underline a long section, the underlining loses its function of pointing out what's *especially* important.

Examples

This is some regular text.

This is some **bold** text.

This text is *italic*.

This text is <u>underlined</u>.

Don't use SMALL CAPS.

Don't use ALL CAPS.

Don't combine font attributes.

In online help, use underlined text for <u>links</u> only.

It's hard to read a long italic text. Try it for yourself: Lorem ipsum dolor sit amet, consectetur adipiscing elit. Aenean ac orci quis massa vestibulum mollis. Sed rutrum purus ac felis blandit sed sodales justo auctor. Suspendisse mi lectus, lobortis vitae consectetur vel, facilisis id nisi. Duis nulla ligula, interdum vulputate suscipit id, fringilla sed nisi. Nam vel lacinia libero. Vivamus sagittis sapien in nibh dapibus a fermentum ligula placerat. Pellentesque dolor enim, volutpat quis ultricies quis, vehicula nec nisi. Sed ornare pellentesque odio eget mattis. Ut fermentum pulvinar nisi, at sodales felis fermentum ac.

3.2.3   Which font size and font spacing?

Don't use a poorly readable font size for the sake of design. In user assistance, readability is more important than beauty.

Keep in mind:

- Many people have poor eyesight. If *you* can read something well, this doesn't mean that *others* can also read it well.
- Your document may be read in places or under conditions where reading is more difficult than in your office.
- Audiences that aren't used to reading long texts may be discouraged by small fonts.
- Type smaller than 10 points usually slows down *all* readers.

General recommendations

Adequate font sizes for printed manuals usually are:

- on small paper sizes: body text from 8.5 points up to 10 points, headings up to 20 points
- on large paper sizes such as A4 or Letter: body text from 10 points up to 12 points, headings up to 24 points
- if the manual is likely to be read under difficult conditions: body text up to 14 points

Adequate font sizes for online help usually are:

- 10 points or 11 points for body text
- 14 points for headings

Other factors that influence font size:

- In texts that are mainly used for looking up small pieces of information, eye exhaustion can be neglected. So you can use a smaller font size here. Dictionaries are a typical example of this.
- Also base your decision on the specific font that you're using. Not all fonts of a given font size actually have the same character size. For example, in one font with a font size of 10 points, letters may be 2.5 mm high, whereas in another font with a font size of 10 points, letters may only be 2.0 mm high.

Font size in headings

The font size is the most important way to visualize the hierarchy of headings (other ways are the font weight, color, and space before and after the heading).

- The size runs from biggest to smallest, and the typeface runs from boldest to lightest.

- In a printed manual, use at least a difference of 3 points between the levels of headings (for example, 20, 17, 14 points). Otherwise, readers might be unsure of what level they're looking at.

- In online help, topics are more independent. Here, use the same font size for all topic titles, regardless of where they're located within the hierarchy of the table of contents.

Optimizing font spacing for large font sizes

For font sizes up to 16 points, the default font spacing usually provides the best results.

If you use font sizes that are larger than 16 points (mainly in headings), you often need to decrease the default font spacing for better readability. As a positive side effect, this also makes the headings shorter and reduces the risk of having unwanted line breaks.

If you use a light font color on a dark background (inverted text), increasing font spacing usually also improves readability.

In online help, slightly increased font spacing can improve readability as well, especially if you don't use a font that's optimized for reading on screen.

3.2.4   What line length?

The number of characters that can be printed within one line depends on:

- the page size
- page margins
- the font (see *Which font?* [84])
- the font size (see *Which font size and font spacing?* [91])

All of these settings should work together to achieve an adequate line length.

If lines are too long:

- When reading, moving the eyes isn't enough. Readers must also move their heads.
- You don't have an overview of the whole line. This makes it difficult to go to the beginning of the next line without slipping into the wrong line.

If lines are too short:

- Sentences are frequently interrupted by line breaks, which slows down reading.
- The layout looks amateurish:
  - If you hyphenate, there are many hyphenated words.
  - If you don't hyphenate, there's a lot of unused white space.

Recommendations

Adequate line lengths are roughly in between:

- 40 to 70 characters per line
- 8 to 12 words per line (this value applies to English; most other languages have slightly longer words, so a number of 7 to 11 words is more adequate there)

Use line lengths near the upper limit of 12 words per line if:

- your readers are well-educated
- your text has many long paragraphs
- you use a serif font

Examples

In the following example, the text has an average of approximately 30 characters. This is clearly too short:

> Now test the readability of different line lengths. You will notice that reading is harder when lines are too short or too long. Now let's make things a little harder. Here comes some Latin text: Lorem ipsum dolor sit amet, consectetur adipiscing elit. Ut faucibus dignissim mattis. Nullam ut lobortis augue. Nulla viverra, elit semper gravida tempor, nulla risus luctus tellus, quis iaculis metus sem sed purus.

An average of approximately 60 characters is best:

> Now test the readability of different line lengths. You will notice that reading is harder when lines are too short or too long. Now let's make things a little harder. Here comes some Latin text: Lorem ipsum dolor sit amet, consectetur adipiscing elit. Ut faucibus dignissim mattis. Nullam ut lobortis augue. Nulla viverra, elit semper gravida tempor, nulla risus luctus tellus, quis iaculis metus sem sed purus. Nullam porttitor sagittis interdum. Cras porta lobortis neque, a pretium libero suscipit sed. Pellentesque habitant morbi tristique senectus et netus et malesuada fames ac turpis egestas. Donec libero sapien, gravida id interdum ut, egestas at erat. Sed non turpis erat. Morbi tristique scelerisque ultricies.

An average of approximately 75 characters is already too long:

> Now test the readability of different line lengths. You will notice that reading is harder when lines are too short or too long. Now let's make things a little harder. Here comes some Latin text: Lorem ipsum dolor sit amet, consectetur adipiscing elit. Ut faucibus dignissim mattis. Nullam ut lobortis augue. Nulla viverra, elit semper gravida tempor, nulla risus luctus tellus, quis iaculis metus sem sed purus. Nullam porttitor sagittis interdum. Cras porta lobortis neque, a pretium libero suscipit sed. Pellentesque habitant morbi tristique senectus et netus et malesuada fames ac turpis egestas. Donec libero sapien, gravida id interdum ut, egestas at erat. Sed non turpis erat. Morbi tristique scelerisque ultricies.

4 Writing

Everyone can write. However, not everyone can write so that everyone understands.

Characteristics of user-friendly style

Good user assistance is:

- correct and unambiguous
- written in a way that makes it easy to mentally process the given information
- written in a way that makes it easy to act upon the given instructions
- written in a way that makes it easy to remember the given information

Levels of writing involved

Creating clear, user-friendly documents involves all levels of writing. It starts with how you organize the information within a topic. It continues with how you structure paragraphs and build sentences. It ends with the choice of words. For details about each level, see the following sections:

- *Writing in general* 97
  Summarizes the key principles that you should follow on *all* levels of writing.

- *Writing sections* 123
  Shows how you should organize the given information into paragraphs, how to add subheadings, and what to bear in mind when writing specific information types such as procedures or warnings.

- *Writing sentences* 151
  Shows how to build sentences that are grammatically simple, clear, and easy to understand.

- *Writing words* 171
  Shows how to choose words that add clarity rather than complexity.

- *FAQ: Grammar and word choice* 201
  Makes you aware of frequent grammar and word choice problems, such as the difference between the words *that* and *which*, or the correct use of the words *safety* and *security*.

4.1   Writing in general

Write for the reader; don't write for your ego. Simpler is better.

Which style to follow?

Make reading your documents a positive experience. Write in a way that's:

- accurate and objective
- informative and helpful
- respectful and friendly
- positive and reassuring

General writing principles

The key overall principles when writing user assistance are:

- *Keep it simple and stupid* 98
- *Always start with the main point* 100
- *Talk to the reader* 102
- *Be specific* 105
- *Be concise* 109
- *Be consistent* 111
- *Be parallel* 113
- *Use the present tense* 117
- *Use the active voice* 118
- *Don't say "please"* 121

4.1.1   Keep it simple and stupid

Forget what you learned at school.

You're NOT writing an essay. You DON'T have to impress your teacher.

Provide information that EVERYBODY can understand—even readers who:

- don't speak the document's language as their first language
- aren't sitting in a silent office but who, for example, are standing in a noisy production hall
- don't have much time
- are frustrated because they didn't succeed without reading the manual

So **keep** **it** **s**imple and **s**tupid (KISS principle):

- Write short sentences.
- Use simple grammar.
- Use simple words.

Plain language is NOT evidence of poor education. Plain language is the foundation of clear user assistance.

✘ No: *If you want to exert influence on the contents of a document, access the submenu item **Edit** in the **File** menu after having opened the document file successfully.*

✔ Yes: *To edit a document:*

1. Open the document file.

*2. Choose **File** > **Edit**.*

✘ No: *Congratulations for buying this sophisticated, highly effective phone, which has been designed with your most vital communication needs in mind.*

✔ Yes: *You can use this phone to make phone calls.*

✔ Top: Leave out this sentence completely because it doesn't provide *any* useful information.

Can you measure simplicity?

Linguists have developed a number of indexes that attempt to measure the degree of complexity of a text. Some text editors have built-in functions to calculate these indices.

Don't take these indices too seriously. A comprehensibility index provides a rough estimate, but comprehensibility is determined by many more factors than average word length and average sentence length.

> **ⓘ Important:** Don't aim for a specific index value that's said to be "adequate" for the educational level of your audience. User instructions can't be too simple. Always aim for *maximum simplicity*. If you feel that your document might look too trivial for your audience, the document probably isn't too simple but too detailed. Try to identify things that you can omit.

Can you use software to guarantee simplicity?

On the market, there are a number of programs that can make suggestions about how to simplify a text. If you have access to one of these programs, go ahead and use it. Most of these programs can give you valuable feedback, however none of them can replace a human editor.

Also, don't forget that it's YOU who must structure and write clearly in the first place. If you don't, neither software nor a human editor will have much of a chance to improve your text.

4.1.2 Always start with the main point

On all levels, provide the key message as soon as possible. Place the main points:

- in the first topic of a section
- on top of the page or screen
- at the beginning of a subsection rather than in the middle
- at the beginning of a paragraph rather than in the middle
- at the beginning of a sentence rather than in the middle
- in the first table column
- in the first table row
- on the left side of a figure (in languages that are read from left to right)

The position in front is the most prominent position:

- Readers assume that what comes right at the beginning is more important than what comes somewhere else.
- Readers remember better what comes at the beginning than what comes somewhere else.
- When readers skim a text, what comes at the beginning is easier for them to find than what comes somewhere else.
- Readers who don't read the full topic at least read the key message at the beginning.

How to identify the main point

It's your job as the author to decide what's most important and thus what becomes the main point. Often, the main point is:

- what most users need to know
- what users need to know early
- what's not optional
- what may cause an error, damage, injury, or death
- what's a prerequisite for an action
- what users must find or do first

(DE)

Innerhalb eines Satzes bietet die deutsche Sprache besonders viele Freiheiten um bestimmte Dinge gezielt zu positionieren.

Im Englischen steht beispielsweise das Subjekt immer vor dem Prädikat, sowohl im Hauptsatz als auch in Nebensätzen (SPO-Regel). Im Deutschen kann das Subjekt den Platz jedoch mit dem Objekt tauschen.

✔ **Ja:**   *Sie können den Nippel durch die Lasche ziehen.*

(Hier liegt die Betonung auf *können.*)

✔ **Ja:**   *Den Nippel können Sie durch die Lasche ziehen.*

(Hier liegt die Betonung auf *Nippel.*)

Treffen mehrere Substantive aufeinander, werden Sätze oft mehrdeutig. Aus Gewohnheit tendieren die meisten Leser allerdings dazu, den zuerst genannten Ausdruck als Subjekt aufzufassen. Bauen Sie daher Ihre Sätze ebenfalls entsprechend diesem Schema auf.

Formulieren Sie einen Satz um, wenn die Möglichkeit einer Fehlinterpretation besteht.

✘ **Nein:**   *Die Beschreibung der Schnittstelle enthält das Kapitel „Port Interface".*

(Unklar: Wer enthält wen?)

✔ **Ja:**   *Das Kapitel „Port Interface" enthält die Beschreibung der Schnittstelle.*

✔ **Top:**   *Die Beschreibung der Schnittstelle finden Sie im Kapitel „Port Interface".*

oder:

Die Beschreibung der Schnittstelle enthält unter anderem auch das Kapitel „Port Interface".

4.1.3   Talk to the reader

1 Talk to the reader directly ("You can ...."). Talking directly to the reader increases attention and avoids ambiguity (see also *Use the active voice* 118).

Don't use the passive voice ("... can be done.").

Don't talk about the user ("Users can ...."). Don't use phrases with "one" ("One can ....").

Don't be afraid of giving commands. It's your job to tell your readers clearly what to do.

Write as though you were talking to your readers in a friendly, straightforward way (conversational style). Keep it simple, make short sentences, and use the same short, everyday words that you use when talking to co-workers.

When giving recommendations, it's acceptable to use *we*. Don't use the passive voice to avoid the editorial *we*. Often, however, the best solution is to create a sentence with *you*.

Exceptions:

2 If you're writing for developers or for administrators, use second person to refer to your reader (the developer or administrator), but use third person to refer to the reader's end user.

3 In error messages and troubleshooting information, it can be more polite to use a passive construction rather than to tell users right away that a problem is their own fault.

4 In tutorials, a passive construction is sometimes appropriate to distinguish general information from a prompt to act.

1

✖ **No:**   *The button must be pressed.*

✖ **No:**   *The button must be pressed by the user.*

✖ **No:**   *Users must press the button.*

✖ **No:**   *One must press the button.*

✔ **Yes:**   **Press the button.**

✖ **No:**   *The Print dialog provides the possibility to change the printer settings.*

✔ **Yes:**   **In the Print dialog, you can change the printer settings.**

✖ **No:**   *In this section, the installation of the program will be shown.*

✔ **Yes:** *In this section, we show you how to install the program.*

✔ **Top:** *In this section, you'll learn how to install the program.*

✖ **No:** *It's recommended to use a shielded cable.*

✔ **Yes:** *We recommend using a shielded cable.*

✔ **Top:** *For best performance, use a shielded cable.*

2

✖ **No:** *Administrators can reset passwords so that users are able to create a new password.*

✖ **No:** *Passwords can be reset so that users are able to create a new password.*

✔ **Yes:** *You can reset passwords so that users are able to create a new password.*

3

✖ **No:** *You've made a serious mistake. Next time, read the manual before you try this.*

✔ **Yes:** *A mistake has been made. More information can be found in the manual.*

4

✖ **No:** *To make a phone call, type the telephone number. Now you try it: Type a friend's telephone number.*

✔ **Yes:** *Phone calls are made by typing the telephone number. Now you try it: Type a friend's telephone number.*

(DE)

Verwenden Sie im Deutschen die direkte Anrede mit *Sie*.

Vermeiden Sie Satzkonstrukte mit *man* oder *es*.

Sprechen Sie immer klar aus, wer gemeint ist und wer handelt.

✖ **Nein:** *Man kann das Passwort auch ändern.*

✖ **Nein:** *Es ist auch möglich, das Passwort zu ändern.*

✖ **Nein:** *Das Passwort kann auch geändert werden.*

✖ **Nein:** *Das Passwort lässt sich auch ändern.*

✘ Nein: *Das Passwort ist änderbar.*

✔ Ja: *Sie können Ihr Passwort auch ändern.*

oder:

Ihr System-Administrator kann Ihr Passwort bei Bedarf für Sie ändern.

✘ Nein: *Den Bildschirm reinigt man am besten mit klarem Wasser.*

✘ Nein: *Der Bildschirm wird am besten mit klarem Wasser gereinigt.*

✔ Ja: *Den Bildschirm reinigen Sie am besten mit klarem Wasser.*

✔ Ja: *Reinigen Sie den Bildschirm am besten mit klarem Wasser.*

Telegrammstil in Handlungsanleitungen

In Handlungsanweisungen ist im Deutschen alternativ auch ein Telegrammstil im Imperativ (Befehlsform) möglich (siehe *Writing procedures* 128).

4.1.4 Be specific

When reading instructions, users are looking for clear answers.

- Don't be vague.
- Don't be ambiguous.

You know the product that you're describing—readers don't. What's clear to *you* may not be clear at all to *your audience*. Also, bear in mind that most readers don't read manuals from start to finish, so they only see a small part of the whole.

Unclear or ambiguous text has a serious impact on the perceived quality of your document:

- If readers *do* notice that a phrase is unclear or ambiguous, this results in uncertainty. Ambiguous texts don't inspire confidence.
- If readers *don't* notice that a phrase is unclear or ambiguous (which often happens), this may result in misunderstanding and failure.
- If an unclear or ambiguous phrase goes unnoticed by translators, translated versions of your document may be plain wrong. In this case, *all* readers who read the translated version get the wrong information.

The key rule on how to be specific is to avoid all sorts of vague terms.

Note:
Being specific is more important than being concise (see *Be concise* 109).
Don't write ambiguous sentences because you want to make them as short as possible. If necessary, don't hesitate to repeat a word, or add a syntactic cue (see *Feel free to repeat a word* 160 and *Add syntactic cues* 163).

✘ **No:** *If you've filled in all fields correctly, the results window should appear.*

✔ **Yes:** *If you've filled in all fields correctly, the results window appears.*

(If you doubt that your product works as intended, don't show your doubt to the reader. Always describe the intended behavior or usual condition.)

✘ **No:** *The action should be finished quickly.*

✔ **Yes:** *You need to finish the action within one minute.*

✘ **No:** *The program can also import a number of other formats.*

✘ No: *The program can also import Word files, etc.*

✔ Yes: **The program can also import Word files, PDF files, and XML files.**

✘ No: *If necessary, turn the lights on.*

✔ Yes: **If it's dark and the lights are off, turn them on.**

✘ No: *between 7 and 11*

(Unclear: Are 7 and 11 included?)

✔ Yes: *from 7 through 11*

✘ No: *The file set includes common files that are used in many web applications.*

✔ Yes: **The file set includes files that are shared between web applications on the web server.**

or:

The file set includes files that many web applications typically use.

✘ No: *We will successfully install your washing machine.*

✔ Yes: **We will unpack your washing machine, set it up, connect it, and get it working.**

✘ No: *Press any key to continue.*

✔ Yes: **Press [Enter] to continue.**

(Note: Name a specific key even if other keys will do the same job. For users, it's faster to look for the [Enter] key than having to choose a key by themselves. In addition, it prevents any feeling of uncertainty.)

Typical examples of vague terms

- *and so on*
- *and/or*
- *can*
- *corresponding*
- *etc.*
- *may*
- *maybe*
- *object*

- *ought to*
- *quite*
- *rather*
- *respectively*
- *should*
- *some*

 (DE)

Problemwort „beziehungsweise"

Im Deutschen ist eine häufige Quelle für schwammige Sätze das Wort „beziehungsweise". Oft ist nur schwer oder gar nicht nachvollziehbar, worin die „Beziehung" besteht.

- Entweder meinen Sie *und*. Dann sagen Sie das auch so.
- Oder Sie meinen *oder*. Dann sagen Sie das auch so.
- Oder Sie meinen *genauer gesagt*. Dann sagen Sie es genauer.
- Oder Sie meinen eigentlich gar nichts, und *beziehungsweise* ist nur ein Füllwort. Dann lassen Sie es weg.

✘ **Nein:** *Füllen Sie Benzin und Wasser in den Tank bzw. in den Vorratsbehälter der Wischanlage.*

✔ **Ja:** *Füllen Sie Benzin in den Tank, und füllen Sie Wasser in den Vorratsbehälter der Wischanlage.*

✔ **Top:** *1. Füllen Sie Benzin in den Tank.*
2. Füllen Sie Wasser in den Vorratsbehälter der Wischanlage.

Hinweis:
Im Englischen gibt es das Wort *beziehungsweise* in der Form wie im Deutschen nicht. Das englische Wort *respectively* bedeutet *in the given order* und setzt eine ganz bestimmte Satzstruktur voraus. Beispiel: „Variables N and M take on values 1 and 2, respectively."

Typische Beispiele für vage Ausdrücke im Deutschen

- *bzw.*
- *einige*
- *entsprechende*
- *es*
- *etc.*
- *eventuell*

- *gegebenenfalls*
- *können*
- *man*
- *Objekt*
- *sollen*
- *und/oder*
- *usw.*
- *ziemlich*

✖ **Nein:** *Auch wenn Sie sich selbst unsicher sind, empfiehlt es sich nicht, gegebenenfalls entsprechend vage Ausdrücke zu verwenden.*

✔ **Ja:** **Verwenden Sie keine vagen Ausdrücke, auch dann nicht, wenn Sie sich selbst unsicher sind.**

✖ **Nein:** *Sichern Sie Ihre Daten täglich. Am Wochenende können Sie gegebenenfalls darauf verzichten.*

✔ **Ja:** **Sichern Sie Ihre Daten nach jedem Arbeitstag.**

4.1.5 Be concise

Omit all words and syllables that are nothing but empty calories.

Every word and character saved is a step toward more clarity. The only exception to this rule is: Don't be concise at the expense of clarity. If you need more words to be more specific or to avoid ambiguity, go ahead and include them (see *Add syntactic cues* 163, *Be clear about what you're referring to* 165, and *Feel free to repeat a word* 160).

The key to avoiding empty calories in your documents is to be aware why you may be tempted to add them:

- When you aren't sure about the facts that you describe, you might be tempted to conceal your uncertainty by adding something vague.
- You might be tempted to impress your readers with your sophisticated language skills or with your profound domain knowledge.
- You might be tempted to impress your boss with the number of pages that you've produced.
- You might be tempted to impress customers with the number of pages because you think that a big manual makes your product look like it's worth the money.
- You just don't care and write down something quickly because you don't like writing manuals and want to complete this task as soon as possible.

Resist these temptations.

✖ **No:** *The program can handle the following four file formats: A, B, C, and D.*

✔ **Yes:** *The program can handle the file formats A, B, C, and D.*

(In this sentence, the relevant fact is which file formats the program supports. The number of formats ("four") is irrelevant, so leave it out. Also you can leave out the phrase "the following" without any loss of information.)

✖ **No:** *The new car is faster and will break down less often.*

✔ **Yes:** *The new car is faster and more reliable.*

✖ **No:** *The cable is about 10 meters in length.*

✔ **Yes:** *The cable is 10 meters long.*

✖ **No:** *It has a rectangular shape.*

✔ Yes: *It's rectangular.*

✘ No: *You should have some experience within a Unix environment.*
✔ Yes: **You need to have some Unix experience.**

✘ No: *If you're a user who has experience in this field, use expert mode.*
✔ Yes: **If you're an experienced user, use expert mode.**

✘ No: *The program isn't able to print.*
✔ Yes: **The program can't print.**

✘ No: *It's necessary to enter a value.*
✘ No: *You're required to enter a value.*
✔ Yes: **You must enter a value.**

✘ No: *You can format the table by means of the toolbar.*
✔ Yes: **To format the table, use the toolbar.**

✘ No: *In order to print the file, choose the menu command File > Print.*
✔ Yes: **To print the file, choose File > Print.**

✘ No: *It takes a longer period of time to write a user manual than to read it.*
✔ Yes: **It takes longer to write a user manual than to read it.**

■ (DE)

Einige Beispiele aus dem Deutschen:

✘ Nein: *Im Rahmen der technischen Weiterentwicklung konnte die Motorleistung um 5 kW gesteigert werden.*

✔ Ja: **Die Motorleistung wurde um 5 kW gesteigert.**
✔ Top: **Der Motor leistet jetzt 55 kW statt zuvor 50 kW.**

✘ Nein: *Wählen Sie den Kalendermonat.*
✔ Ja: **Wählen Sie den Monat.**

✘ Nein: *Stellen Sie Ihren Sitz in eine senkrechte Position.*
✔ Ja: **Stellen Sie Ihren Sitz senkrecht.**

4.1.6   Be consistent

A consistent document is easy to read. The reader can fully focus on the content.

In addition, a consistent document makes a professional impression, which increases your credibility and builds up confidence.

Document consistency is a result of:

- consistent design and formatting
- parallel structures and phrases (see *Be parallel* 113)
- consistent spelling, punctuation, and choice of words (see *Always use the same terms)* 182

Aim for consistency not only within each document but also among all documents.

✖ No: *Documentation for product A:*

- *Setting up Product A*
- *First Steps with Product A*
- *Using Product A*
- *Product A Technical Reference*

Documentation for product B:

- *Installing Product B*
- *Getting Started with Product B*
- *Product B User's Guide*
- *Product B Developer's Guide*

✔ Yes: *Documentation for product A:*

- *Product A Installation Guide*
- *Product A Getting Started Guide*
- *Product A User's Guide*
- *Product A Developer's Guide*

Documentation for product B:

- *Product B Installation Guide*
- *Product B Getting Started Guide*

- *Product B User's Guide*
- *Product B Developer's Guide*

Tips for obtaining consistency

Consistency can be hard to achieve, especially when more than one author works on the same document. But even if you're the only author, it's often difficult to remain consistent in the long run.

To remain consistent:

- Create common document templates that all authors are obliged to use.
- Keep track of your preferences in a terminology and preferences list.

Sample structure of a simple terminology list

In many cases, you don't have to use a dedicated terminology database or terminology management system to achieve a consistent use of terms. Even a simple, short terminology list written in any spreadsheet program can work wonders.

For example, a basic terminology list could look like this:

Term to use	Terms NOT to use	Comments
computer	client device machine PC unit workstation	When referring to computer networks, it's OK to use *client* in this particular context.
...	...	...
...	...	...

Don't use the terms listed in the "Terms NOT to use" in your visible texts, but do add them as index keywords to support readers who use them.

Tip:
Add all terms that you *don't* want to use to the exclusions list of your spelling checker so that the spelling checker marks them as *wrong*. This is an inexpensive and effective way to identify undesirable terms automatically, especially if your spelling checker checks the spelling as you type (live spelling check).

4.1.7 Be parallel

In technical writing, repetitive structures aren't a sign of weak style but enhance readability.

Parallel structures make the content more predictable. Readers don't have to mentally process a new structure but can, instead, attend to the words alone.

When possible, structures should be parallel:

- between sentences
- within a sentence

Parallelism is especially important in phrases with *and* and *or*.

Parallelism in headings

Try to keep all headings within a chapter, section, or other unit grammatically parallel, especially those on the same level. If it doesn't make sense to keep all headings parallel, try to keep at least as many subsequent headings as parallel as possible.

✖ No: *Washing of trucks*
       Washing cars
       How can you wash a motorcycle?
       How to wash a bicycle

✖ No: *Washing trucks*
       Washing cars
       Washing a motorcycle
       Washing a bicycle

✖ No: *Washing trucks*
       Cleaning cars
       Giving a wash to motorcycles
       Cleansing bicycles

✔ Yes: *Washing trucks*
        Washing cars
        Washing motorcycles
        Washing bicycles

Parallelism in lists and tables

Don't mix full sentences with sentence fragments (see also *Writing lists* 138).

✖ No: *In a document, white space is important for the following reasons:*
  - *visual separation of sections*
  - *attention focus*
  - *white space breaks content into smaller chunks*

✔ Yes: *In a document, white space is important because it:*
  - *visually separates sections*
  - *focuses attention*
  - *breaks content into smaller chunks*

Parallelism in procedures

Typically, begin all steps with a verb (see also *Writing procedures* 128).

✖ No: *To print a picture:*
  1. *Open the image file.*
  2. *Choose File > Print > Options.*
  3. *Next, select the option Photo Quality.*
  4. *Finally, the Print button must be clicked.*

✔ Yes: *To print a picture:*
  1. *Open the image file.*
  2. *Choose the menu command File > Print > Options.*
  3. *Select the option Photo Quality.*
  4. *Click the Print button.*

(DE)

Im Deutschen haben Sie zusätzlich die Wahl, sich für oder gegen eine Formulierung im Infinitiv zu entscheiden.

✖ Nein: *Um ein Rad zu entfernen:*
  1. *Parken Sie das Fahrzeug auf einer waagerechten Fläche.*
  2. *Bitte Feststellbremse gut anziehen.*
  3. *Nun die Radmuttern geringfügig lockern, jedoch noch nicht entfernen.*
  4. *Lösen und entfernen Sie die Radmuttern erst, nachdem Sie das Fahrzeug mit Hilfe des Wagenhebers angehoben haben.*
  5. *Fertig! Jetzt können Sie das Rad abnehmen.*

✔ Ja: *Um ein Rad zu entfernen:*
  1. *Parken Sie das Fahrzeug auf einer waagerechten Fläche.*
  2. *Ziehen Sie die Feststellbremse gut an.*

> *3. Lockern Sie die Radmuttern geringfügig, entfernen diese jedoch noch nicht.*
>
> *4. Heben Sie das Fahrzeug mit Hilfe des Wagenhebers an.*
>
> *5. Lösen und entfernen Sie die Radmuttern ganz.*
>
> *6. Nehmen Sie das Rad ab.*
>
> alternativ:
>
> *Um ein Rad zu entfernen:*
>
> *1.Fahrzeug auf waagerechter Fläche parken.*
>
> *2.Feststellbremse gut anziehen.*
>
> *3.Radmuttern geringfügig lockern, jedoch noch nicht entfernen.*
>
> *4.Fahrzeug mit Wagenheber anheben.*
>
> *5.Radmuttern ganz lösen und entfernen.*
>
> *6.Rad abnehmen.*

Parallelism within sentences

Balance parts of a sentence with their correlating parts (nouns with nouns, prepositional phrases with prepositional phrases, and so on).

Note:
Sometimes you need to repeat a word. This is perfectly OK (see *Feel free to repeat a word* 160).

✖ **No:** *The program can be used to manage annual reports, budgets, and financial planning.*

✔ **Yes:** *You can use the program for managing annual reports, for budgeting, and for financial planning.*

✔ **Yes:** *You can use the program to manage annual reports, budgets, and financial plans.*

✖ **No:** *Whether at home or when working, DemoProduct helps you.*

✔ **Yes:** *Whether at home or at work, DemoProduct helps you.*

✔ **Yes:** *When at home or when working, DemoProduct helps you.*

✖ **No:** *Your alternatives are to use feature A or using feature B.*

✔ **Yes:** *Your alternatives are using feature A or using feature B.*

✔ **Yes:** *Your alternatives are to use feature A or to use feature B.*

✖ **No:**   *The product can be used by children and adults.*

✔ **Yes:**   *The product can be used by children and by adults.*

✔ **Top:**   *Children and adults can use the product alike.*

(This version is better because it avoids the passive voice.)

✖ **No:**   *The traffic light may be red or green. Green means that you can go, red means that you must stop.*

✔ **Yes:**   *The traffic light may be red or green. Red means that you must stop, green means that you can go.*

(Here, the second sentence uses the same order of colors as the first sentence: first red, and then green.)

4.1.8   Use the present tense

> **1** Write as if you were talking over the reader's shoulder. Write as if the reader was using the application right now. Use the present tense.
>
> Using the present tense implies that things always happen as described. This inspires confidence.
>
> **2** Save the future tense for things that will happen in the future.
>
> The future tense is also appropriate if the product or a particular feature that you describe isn't yet available.

1

✖ **No:**   *Click the Send button. Your email will be sent to the recipient.*

✔ **Yes:**   *Click the **Send** button. The program now sends your email to the recipient.*

✖ **No:**   *The Printer Options window will appear.*

✔ **Yes:**   *The **Printer Options** window appears.*

2

✖ **No:**   *You receive an answer to your email within one hour.*

✔ **Yes:**   *You will receive an answer to your email within one hour.*

✖ **No:**   *The upcoming version is able to ....*

✔ **Yes:**   *The upcoming version will be able to ....*

4.1.9 Use the active voice

1 Use the active voice rather than the passive voice.

- The active voice always makes clear whether the reader must act or whether the system acts automatically.
- The active voice is shorter and easier to understand than the passive voice. It directly tells the users what to do. This is particularly important in warning messages.

Note:
The rule to avoid the passive voice sometimes conflicts with the rule to put the more important information before the less important information within a sentence. If this happens, use the active voice anyway or rephrase the sentence.

Exceptions:

2 You can use the passive voice intentionally if you want to avoid pointing a finger at someone. In error messages and troubleshooting information, using the passive voice can be a polite way to tell users that a problem is their own fault.

3 In tutorials, a passive construction is sometimes appropriate to distinguish general information from a prompt to act.

1

✘ No: *The active voice is to be used.*
✔ Yes: *Use the active voice.*

✘ No: *To enter the room, the door must be unlocked.*
✘ No: *The door must be unlocked to enter the room.*
✘ No: *If users want to enter the room, they must unlock the door.*
✘ No: *One must unlock the door before the room can be entered.*
✘ No: *In order to enter the room, it's necessary that the door gets unlocked.*
✔ Yes: *To enter the room, unlock the door.*

✘ No: *It is recommended to ....*
✔ Yes: *We recommend ....*

✘ No: *Now a backup can be made.*

✔ Yes: *Now you can make a backup.*

or:

Now the system automatically makes a backup.

✘ No: *More information can be found in the appendix.*

✔ Yes: *You can find more information in the appendix.*

2

✘ No: *You've made a serious mistake. Next time, read the manual before you try this.*

✔ Yes: *A mistake has been made. More information can be found in the manual.*

3

✘ No: *To make a phone call, type the telephone number. Now you try it: Type a friend's telephone number.*

✔ Yes: *Phone calls are made by typing the telephone number. Now you try it: Type a friend's telephone number.*

▬ (DE)

Typische Beispiele in deutscher Sprache:

✘ Nein: *Nach dem Herunterfahren wird die Anlage ausgeschaltet.*

✔ Ja: *Nach dem Herunterfahren schaltet sich die Anlage automatisch aus.*

oder:

Schalten Sie nach dem Herunterfahren die Anlage aus.

✘ Nein: *Nachdem die Adresse eingegeben wurde, kann die Nachricht gesendet werden.*

✔ Ja: *Nachdem Sie die Adresse eingegeben haben, können Sie die Nachricht senden.*

oder:

Nachdem Sie die Adresse eingegeben haben, versendet das Programm selbsttätig die Nachricht.

✘ Nein: *Der Status kann nicht geprüft werden.*

✔ Ja: *Das System kann den Status nicht prüfen.*

✖ Nein: *Technische Daten können dem Anhang entnommen werden.*

✔ Ja: **Technische Daten finden Sie im Anhang.**

✖ Nein: *Die gültigen Sicherheitsvorschriften sind einzuhalten.*

✔ Ja: **Halten Sie die gültigen Sicherheitsvorschriften ein.**

✖ Nein: *Nach 10.000 km ist das Öl zu wechseln.*

✔ Ja: **Wechseln Sie nach 10.000 km das Öl.**

4.1.10   Don't say "please"

> **1** When you're giving instructions, you're *not* asking readers for a favor.
>
> For this reason, using the word *please* is inappropriate and just bloats your text. (Also, where would you stop? Would you say *please* in every sentence? Once in a paragraph? Once in a topic? It just doesn't make sense.)
>
> Using the word *please* can even be misleading because it implies that an action is optional, which in many cases probably isn't the case.
>
> **2** The only time when it's appropriate to say *please* is when you're apologizing for a problem, or when you're actually asking for a favor.

1

✘ **No:**   *Please insert the program CD into the disc drive.*

✔ **Yes:**   *Insert the program CD into the disc drive.*

2

✔ **Yes:**   *If you get an error message, please contact support.*

✔ **Yes:**   *Can we improve this manual? Please send your feedback to feedback@yourdomain.com.*

4.2 Writing sections

Don't mix information. Label each section with a descriptive subheading that clearly communicates what the section is about. Within the section, only cover what the subheading indicates.

Rules for writing on the section level

- *Don't mix subjects* `124`
- *Add labels (subheadings)* `126`
- *Writing procedures* `128`
- *Writing lists* `138`
- *Writing warnings* `142`
- *Writing cross-references and links* `146`

4.2.1 Don't mix subjects

Long, continuous text can be extremely hard to read, especially when it covers various subjects. Long sections of continuous text also make it almost impossible to skim a text for specific information.

1 Don't mix various subjects within one paragraph. If there's a new subject, start a new paragraph.

It's perfectly OK if sometimes there's only one sentence within a paragraph. Don't combine paragraphs or add needless text only for the sake of building a "complete" paragraph.

As a general rule, start a new paragraph for each new idea, but don't start a new paragraph for each new sentence. A good paragraph length is about 2 to 5 lines.

2 Don't mix various subjects within one sentence. If there's a new subject, start a new sentence.

In procedures, don't describe more than 2 steps in the same sentence (see *Writing procedures* 128).

In warnings, only mention one cause of the hazard per sentence (see *Writing warnings* 142).

1

✘ No: *You can use the camera for taking pictures and for making short movies. Possible resolutions for pictures are a x b pixels and c x d pixels. Possible resolutions for movies are e x f pixels and g x h pixels. Pictures can be saved in JPG format; movies can be saved in AVI format.*

(Several subjects are mixed within this paragraph.)

✔ Yes: *You can use the camera to take pictures and to make short movies.*

Possible resolutions for pictures are a x b pixels and c x d pixels. Possible resolutions for movies are e x f pixels and g x h pixels.

Pictures can be saved in JPG format; movies can be saved in AVI format.

(Here, the subjects are separated into sections: what you can do with the camera (first paragraph), the possible resolutions (second paragraph), and the formats (third paragraph).)

✔ Yes: *You can use the camera to take pictures and to make short movies.*

Possible resolutions for pictures are a x b pixels and c x d pixels. Pictures can be saved in JPG format.

Possible resolutions for movies are e x f pixels and g x h pixels. Movies can be saved in AVI format.

(Here, the information is rearranged. The idea of the second paragraph is "pictures"; the idea of the third paragraph is "movies.")

✔ **Yes:** *Camera A has been designed for children. Camera B has been designed for adults.*

(There's no new paragraph because the direct comparison of both models is one common subject.)

✔ **Yes:** *Camera A has been designed for children.*

You can use the camera to:

- *take pictures*
- *make short movies*

Each function can be activated with a single key.

(Here, there are separate paragraphs because there are several independent ideas: the fact that the camera is for children, the different things you can do with the camera, and the fact that each function can be activated with a single key.)

2

✘ **No:** *You can use the text import function, which can import PDF files, Word files, and XML files, to take over texts that were written with another program.*

✔ **Yes:** *You can use the text import function to take over texts that were written with another program. You can import PDF files, Word files, and XML files.*

4.2.2   Add labels (subheadings)

Start each new section with a label that clearly communicates what the section is about.

In a printed document, labels often appear as subheadings or as subtitles in a margin column. In online help, labels often appear as subtitles as well, or they are clickable headings of expandable sections (toggles).

Readers benefit from the labels because the labels work as landmarks that enable them to skim a text for specific information and to decide beforehand whether to invest the time in reading a particular section.

As the author, you also benefit from the labels:

- Labels give you a constant overview of the structure of your document. You can quickly see where the best place is to add new information.

- Labels minimize the risk of mixing subjects (see *Don't mix subjects* 124). When writing, you can constantly check your text against the label. If the text doesn't fit under the label, this indicates that you should put the information somewhere else.

How to phrase a label

You can either use a short sentence or statement as label text, or you can write the label like a heading.

- Make sure that the label text is concise and easy to read.

- Make the label meaningful. The label should clearly communicate:

  - what the section covers (the subject)

  - what kind of information readers can expect (concept, task, or reference)

  - what level of detail is given (basic information for beginners or details for advanced users)

If the surrounding topic doesn't mix information types but is clearly either a Concept topic, a Task topic, or a Reference topic, labels can be very short and don't have to communicate the information type again. For example, if the surrounding topic is clearly a Task topic, it's evident that subsections also cover tasks.

If the surrounding topic mixes various information types (generally not recommended), you need to find labels that communicate the information type of each section as well. For example, you could use phrases such as "How to print ..." or "Printing ..." to indicate that a section contains a procedure; compare Structuring: *Find meaningful headings* 37 .

How many sentences and paragraphs should go into one labeled section?

There's no general rule about how many labels you should add. A good average is somewhere between 2 and 7, but this largely depends on the subject.

As rules of thumb:

- Never mix different subjects within one section under the same label. If there's a new subject, start a new section.

- When in doubt, it's better to start a new section with an additional label than to have a section that's longer than half a page.

- A section may contain a single paragraph or multiple paragraphs. The paragraphs may all be of the same type (such as procedures) or of different types (such as body text, a procedure, and a warning).

Establish a consistent labeling scheme

When possible, establish a consistent way of structuring your topics and of labeling the sections within these topics. This adds consistency and parallelism (see *Be consistent* 111 and *Be parallel* 113).

For example, you could:

- always use the sections "Requirements," "Steps," and "Results" in Task topics

- always use the sections "Purpose," "Syntax," "Input," "Output," and "Parameters" in a particular kind of Reference topic

Note that it depends on your specific product and use case which labels work best.

4.2.3 Writing procedures

Procedures are the core part of user assistance. A procedure is a description of the steps that users have to follow to complete a specific task.

- A procedure typically begins with an introductory phrase that states the goal of the procedure and ends with a colon. Sometimes the introductory phrase can also be a statement, which ends with a colon or period.

- Steps are numbered. They consist of action statements, descriptions of the system's response, and related information that enable users to execute the procedure.

- The procedure typically ends with a brief description or picture of the result.

✔ **Yes:**   *To set the alarm time:*

1. *Hold down button A for at least three seconds.*
   The time display now blinks and shows the currently set alarm time.

2. *Press the arrow buttons repeatedly to change the alarm time.*

3. *Press button B to store the new alarm time.*
   The new alarm time now appears at the bottom of the display after the chime symbol.

✔ **Yes:**   *Writing a manual involves three major tasks:*

1. *Planning the structure.*

2. *Designing the template.*

3. *Writing the texts.*

(Strictly speaking, this is not a classical procedure but a process description. The introductory phrase is a statement, although it also states the goal (writing a manual). The steps aren't written in an imperative form.)

Basic rules

Never use body text for procedures. Always use numbered steps.

Don't use words to describe the succession of steps, such as "first," "next," "in the following step," and so on.

Don't merge multiple actions into one step.

✖ **No:** *To change the brightness of the screen, you first need to press the red button. The value of the current setting then appears on screen. Next, press the arrow keys to change the brightness. When you've finished, finally press the green button. The new brightness value is now set.*

✖ **No:** *To change the brightness of the screen:*

1. First, press the red button. The value of the current setting now appears on screen. Next, press the arrow keys to change the brightness. When you've finished, continue with step 2.

2. Finally, press the green button.
Congratulations! The new brightness value is now set.

✔ **Yes:** **To change the brightness of the screen:**

1. Press the red button.
The value of the current setting appears on screen.

2. Press the arrow keys to change the brightness.

3. Press the green button.
The new brightness value is now set.

Talk to readers clearly and directly and use the imperative.

✖ **No:** *The next step consists of pressing the Stop button.*

✖ **No:** *One must press the Stop button now.*

✖ **No:** *Now the Stop button should be pressed.*

✖ **No:** *Now it's necessary to press the Stop button.*

✖ **No:** *Now you need to press the Stop button.*

✖ **No:** *Now you must press the Stop button.*

✖ **No:** *Now, please press the Stop button.*

✔ **Yes:** **Press the Stop button.**

Describe only the simplest, most common way of doing things. When documenting software, describe keyboard shortcuts in a separate reference topic of shortcuts.

✖ **No:** *Click File > Save. Alternatively, you can also press [Ctrl]+[S].*

✔ **Yes:** *Click **File** > **Save**.*

If a procedure involves any hazard, add a warning *directly before the step that involves the hazard* (see *Writing warnings* 142). Don't add the warning after the step.

If a warning relates to the procedure as a whole, add the warning at the beginning of the procedure.

Keep procedures as short as possible. If a procedure has more than 7 to 10 steps, consider splitting the procedure. If it's not possible to split the procedure, add labels (subheadings) that group steps into subtasks or that mark certain milestones in the procedure.

Mention all prerequisites

Don't assume that the correct conditions are already in place.

- With products that involve navigation through menus, don't assume that users already are in the right place to start a procedure. They could be *anywhere* within the on-screen menu when reading your instructions.

- In a service manual, don't assume that users have already disassembled certain components or have already completed other preparatory steps.

- Bear in mind that users may have done something that will prevent the procedure from working.

Consequently, clearly state all prerequisites at the beginning of the procedure. In addition, if any special tools are required, list these tools so that users can fetch them in advance.

Often, the best way of stating the prerequisites is to make the preparation the first step of the procedure ("1. Make sure that you have … available.", "1. Prepare ….").

Alternatively, you can formulate the prerequisites as a condition that you put before the steps ("You must have … before …."). If you have a large number of prerequisites, put them into a table or list.

If you have various similar procedures, state the prerequisites in a consistent, parallel way. For example, always use a table, and always use the same structure within this table.

Mention the purpose of a step if this is information is helpful

Don't hesitate to explain the purpose of a step. People are often more willing to execute a particular step and perform it better when they understand why they need to do it. Half a sentence is often enough.

✖ **No:** …

> 5. Fix the bolt with a small piece of adhesive tape.

…

✔ **Yes:** …

> 5. **To prevent the bolt from falling down, temporarily fix it with a small piece of adhesive tape.**

…

Describe the system's responses

To give users the possibility to control the success of their actions, describe not only the users' actions but also what happens as a result of critical actions. Feedback is especially important if many of your readers are inexperienced users.

Sometimes you can embed the information about the response right into a step. If that's not possible, add a line break, and then add a second sentence.

✔ **Yes:**   *1. Click Options to display more fields.*

           2. Select Landscape.

✔ **Yes:**   *1. Click Options.*
             The dialog box expands.

           2. Select Landscape.

If the change of state has important negative implications, include these in a note.

At the end of the procedure, describe or show the final result.

First tell where to act, then tell what to do

Make it easy for readers to follow your instructions step by step. Get your words into the same order that users must follow both mentally and physically.

Match the order of words with the sequence of steps that users must take to identify an object. Always tell users where the action takes place before describing the action to take.

✘ **No:**   *Click Save in the Options window.*

✔ **Yes:**   *In the **Options** window, click **Save**.*

           (Users must first find the **Options** window, then find and click the **Save** button.)

Combining steps

Usually, create one sentence for each step. This breaks down a complex task that's difficult to perform into a number of simple subtasks that are easy to perform. In addition, it helps you to bring the information into the right order.

Avoid "do-this-after-you-have-done-that" statements.

✘ **No:**   *To write a user manual, enter the text into the previously opened word processor, according to the structure of the outline that you've set up before beginning to write.*

✔ **Yes:**   *To write a user manual:*

   1.Set up an outline.

   2.Open your word processor.

   3.Enter the text according to the structure of your outline.

It's OK to combine two steps if:

- both steps are short
- both steps occur in the same place or affect the same component

✔ **Yes:**   *On the **Tools** menu, click **Options**, and then click the **Edit** tab.*

✔ **Yes:**   *Insert the probe holder and turn it clockwise until it snaps into place.*

Note that the word *then* is not a coordinate conjunction and thus cannot correctly join two independent clauses. Always add "and."

✘ **No:**   *On the Tools menu, click Options, then click the Edit tab.*

✔ **Yes:**   *On the **Tools** menu, click **Options**, and then click the **Edit** tab.*

Handling branches and alternatives

Make sure that each numbered step contains at least one action for every reader. For this reason, optional steps shouldn't have their own number.

Always start with the most common (default) action, followed by the alternative action.

For choices within one procedure step, use a list.

✘ **No:**   ...

   7. If you live in Europe, enter EUR.

   8. If you live in America, enter AME.

   9. If you live in any other part of the world, enter GLOB.

   ...

✔ **Yes:**   *...*

   7.Enter your region code:

   - *If you live in Europe, enter EUR.*
   - *If you live in America, enter AME.*
   - *If you live in any other part of the world, enter GLOB.*

   ...

   (Note: This example assumes that the product is mainly sold in Europe, so Europe is the first item in the list.)

Put the if-condition at the beginning of the sentence so that readers for whom the condition isn't fulfilled can skip the rest of the sentence.

✘ No: *Open the cover and check that a color cartridge has been inserted if you want to print in color.*

✔ Yes: *If you want to print in color, open the cover and check that a color cartridge has been inserted.*

Handling loops

When users have to repeat one or more steps multiple times, you often can't avoid having a loop within your procedure.

Always loop to something recognizable, like a step number.

✔ Yes: ...

7. Repeat steps 3 through 6 until ....

...

Note:
Looping to a step number is usually the most user-friendly solution because this clearly identifies the beginning of the loop at a glance. However, bear in mind that this solution is also very error-prone. If you add or remove a step from the procedure later, the step numbers will change. You must then also change the loop information, which you can easily forget to do.

Capitalization and punctuation in procedures

Precede all procedures with colons, regardless of whether the text before the colon is a complete sentence or partial sentence.

Note:
Microsoft recommends that you don't use a colon or any other punctuation here. If you want to create documents that mimic Microsoft user assistance, omit the colon.

Use sentence-style capitalization for each step.

Add a period at the end of each step. Don't use exclamation points.

Alternatives to step-by-step procedures

Sometimes, step-by-step procedures aren't the right means to communicate a task.

Flowcharts:

If a procedure is essentially a decision-making process rather than a straightforward action, consider using a flowchart. A flowchart graphically shows the decision points and the pathways that users must follow through the branching logic of the procedure.

In particular, flow charts are often a good choice for troubleshooting and repair procedures. If the results of the decision tree are recommended actions, you can link to detailed step-by-step procedures from the flow chart.

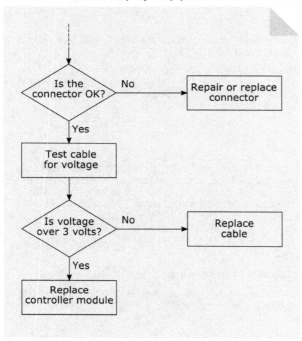

Play script format:

If a task involves various stakeholders, a play script format can be a good choice. Often, such tasks are administrative tasks within an organization.

Each person's actions are listed separately, so each person clearly sees which steps must be done by him or her, and also sees the global context of these actions in the whole procedure.

One column lists the actors; the other column lists the corresponding steps.

Person in charge	Step
Administrator	**1.** The first step.
	2. The second step.
Operator	**3.** The third step.
Technician	**4.** The fourth step.
	5. The fifth step.
	6. The sixth step.
Administrator	**7.** The seventh step.
	8. The eighth step.

 (DE)

Beachten Sie im Deutschen folgende Besonderheiten:

Telegrammstil

Anders als im Englischen lässt sich im Deutschen durch einen (imperativischen) Telegrammstil eine starke Textverkürzung erreichen. Speziell in Dokumenten, in denen Kürze besonders wichtig ist, ist der Telegrammstil daher sehr verbreitet.

Beispiel:

✔ **Ja:** Ausformulierter Stil:

1. Schließen Sie das Antennenkabel an.

2. Schließen Sie das Stromkabel an.

3. Schalten Sie den Fernseher ein.

✔ **Ja:** Telegrammstil:

1. Antennenkabel anschließen.

2. Stromkabel anschließen.

3. Fernseher einschalten.

Vorteile des Telegrammstils:

- kürzer
- durch die Kürze lässt sich oft der Bezug zwischen Bildern und Text besser aufrechterhalten
- die Stellung der Wörter im Satz entspricht automatisch dem Denkmuster „was muss ich nehmen -> was muss ich damit tun?" (Objekt –> Tätigkeit)

Nachteile des Telegrammstils:

- unpersönlicher, da die direkte Anrede fehlt
- kann das Gefühl vermitteln, dass mit dem Produkt keine „vollwertige" Dokumentation mitgeliefert wird, sondern lediglich eine Kurzanleitung

Wägen Sie die Vorteile und Nachteile für Ihren speziellen Anwendungsfall gegeneinander ab. In Branchen mit starkem angelsächsischen Einfluss (z. B. im Software-Bereich) ist fast ausschließlich nur der ausformulierte Stil mit direkter Anrede („Sie") üblich.

Tipp:
Sie können einen Stil auch gezielt nur für einzelne funktionale Elemente oder Dokumente einsetzen. Zum Beispiel können Sie in einer Kurzanleitung den Telegrammstil verwenden, im Benutzerhandbuch jedoch den ausformulieren Stil. Wenn Sie dies konsequent durchhalten, ist das kein Konsistenzbruch.

Formulieren einleitender Sätze

Im Englischen beginnen die einleitenden Sätze zu einer Handlungsanweisung fast immer mit „To ...". Im Deutschen haben Sie ein breiteres Spektrum an Möglichkeiten. Achten Sie jedoch immer darauf, ein Verb zu verwenden. Formulieren Sie einheitlich, so dass die Leser schon am Satzaufbau erkennen, dass eine schrittweise Anleitung folgt.

✖ **Nein:** *Zur Installation von DemoSoft:*

    1. ...

(Hier fehlt ein Verb.)

✖ **Nein:** *Zum Installieren von DemoSoft:*

    1. ...

(Hier fehlt ein Verb.)

✔ **Ja:** *Um DemoSoft zu installieren:*

    1. ...

✔ **Ja:** *So installieren Sie DemoSoft:*

    1. ...

Beginnen Sie einen handlungsanweisenden Satz nur dann mit dem Wort *Wenn*, wenn es sich tatsächlich um eine Auswahlentscheidung handelt. Verwenden Sie *Wenn* nicht als Ersatz für *Um ... zu ...*.

✖ **Nein:** *Wenn Sie die Datei speichern möchten, drücken Sie Strg+S.*

✔ **Ja:** *Um die Datei zu speichern, drücken Sie Strg+S.*

Wenn Sie die Datei unter einem anderen Namen als dem bereits vergebenen Namen speichern möchten, drücken Sie Strg+Shift+S.

Wiederholen des Wortes „Sie"

Wiederholen Sie das Wort *Sie* so häufig wie nötig, sofern dies die Klarheit des Texts erhöht (vgl. *Feel free to repeat a word* | 160 | und *Add syntactic cues* | 163 |).

Meist ist eine Wiederholung des Wortes *Sie* insbesondere dann hilfreich, wenn im zweiten Teilsatz ein anderes Verb steht.

✘ **Nein:** *Klicken Sie auf Datei und zeigen auf Neu.*

✔ **Ja:** *Klicken Sie auf **Datei**, und zeigen Sie dann auf **Neu**.*

4.2.4 Writing lists

Lists present small portions of information in a well-structured way. This helps readers to scan a text quickly and to find information easily.

Lists also help you, the author, because they organize your thoughts.

There's no strict sequence in the items of the list.

List items are bulleted, not numbered. Don't confuse lists with procedures (see *Writing procedures* 126).

Basic rules:

1 If it's not obvious from the heading what a list is about, begin the list with an introductory phrase. The introductory phase can either be a complete sentence or a sentence fragment.

Don't build a list and its introductory phrase as one continuous sentence.

- Don't continue the introductory sentence through the list.
- Don't use words and phrases such as "and," "as well as," or "or" to connect the items of a list.

There are several reasons why you shouldn't build lists that are actually one big sentence:

- You should support readers who don't read the complete text, so it's important that each list item is comprehensible on its own.
- Lists that are actually one big sentence can be very hard to translate into foreign languages.
- Adding items to the list, deleting items from the list, or reversing the order of items requires modifications also in other list items, which is often forgotten.

2 Keep all list items as short as possible. List items may be full sentences or sentence fragments; however, avoid mixing full sentences with sentence fragments within the same list.

Build all list items in a parallel, consistent way (see *Be parallel* 113).

✖ No: *You may use*

- *a CD,*
- *a DVD, or*
- *a USB drive*

to save your data.

✔ **Yes:** *You can save your data on:*

- *CD*
- *DVD*
- *USB drives*

2

✖ **No:** *In a document, white space is important for the following reasons:*

- *visual separation of sections*
- *attention focus*
- *white space breaks content into smaller chunks*

✔ **Yes:** *In a document, white space is important because it:*

- *visually separates sections*
- *focuses attention*
- *breaks content into smaller chunks*

Punctuation and capitalization in lists

Usually, regardless of whether an introductory phrase is a complete sentence or a sentence fragment, place a colon after the introductory phrase. The colon clearly signals the beginning of a list and creates anticipation.

Note:
If the introductory phrase is a full-sentence statement, a colon is sometimes too emphatic. It's OK to use a period then.

✔ **Yes:** *With DemoSoft, you can:*

- *write texts*
- *plan projects*

✔ **Yes:** *You can use DemoSoft to write several types of documents:*

- *letters*
- *reports*
- *books*

If *all* list items are complete sentences, use sentence-style capitalization and punctuation. A list item is considered a complete sentence if, removed from the context of the list, it could stand on its own as a sentence.

✔ **Yes:** *To avoid loss of data:*

- *Make a daily backup copy of all files.*
- *Use the latest security software and update this software regularly.*

If *all* list items are sentence fragments:

- Use lowercase for the initial word of each list item (except for names).

  Many style guides also suggest capitalizing the first word in this case. This is also OK. However, no matter which format you prefer, use it consistently throughout your whole document.
- Don't use periods, semicolons, or commas to end list items if these list items aren't full sentences. Adding punctuation isn't strictly wrong, but it adds unnecessary clutter.
- Don't use a period after the last list item.
- Don't use a period even if a list item completes the introductory clause of the list (not recommended).

✘ **No:** *With DemoSoft, you can:*

- *write texts,*
- *plan projects.*

✔ **Yes:** *With DemoSoft, you can:*

- *write texts*
- *plan projects*

If *some* list items are complete sentences and other list items in the same list are sentence fragments (not recommended):

- Start each list item with a capital letter.
- End each list item with a period.

If essentially all list items are sentence fragments, but one or more of these fragments are followed by a complete sentence, use sentence-style capitalization and punctuation for all list items. As an alternative, you can also use semicolons.

✔ **Yes:** *With DemoSoft, you can:*

- *Write texts. (You must have the extended text module installed to use advanced editing features.)*
- *Plan projects.*

alternatively:

With DemoSoft, you can:

- *write texts; (you must have the extended text module installed to use advanced editing features)*
- *plan projects*

 (DE)

Im Deutschen gelten für Listen dieselben Regeln wie im Englischen.

4.2.5 Writing warnings

Warnings alert users to potential hazards to people or products.

Particular warnings are also often required for legal reasons.

> ℹ **Important:** Comply with the appropriate safety standards, depending on your product and on the countries where your product is sold. These rules have priority over the general recommendations given here. For example, if warnings follow the ANSI Z535.4 and ISO 3864-2 standards, you must add a specific safety symbol, depending on the particular hazard.

It's important not just to tell users what to do or not to do, but to help them understand *why* they should take some particular precautionary measures. Only if users are aware of the reasons for a measure, and of the personal consequences and implications of not following the advice, will they take the warning seriously. For this reason, each warning must provide the following information (in this order):

- Safety alert symbol.
- Signal word that indicates the severity of the hazard: *Caution*, *Warning*, or *Danger* (see the following section on the types of warnings).
- Information on what kind of danger exists, and on where the danger comes from.
- Information in what can happen to the user, to other people, to the product, and to other things.
- Information on how the danger can be avoided altogether, or on how the risk can at least be minimized.

✔ **Yes:** ⚠ *WARNING*
Moving parts can snag and pull.
May cause severe injury.
Do not wear loose clothing or jewelry, and pull back long hair in a hairnet or ponytail while operating the machine.
Do not operate the machine with the guard removed.

ANSI Z535.4 and ISO 3864-2 standard warnings look like this:

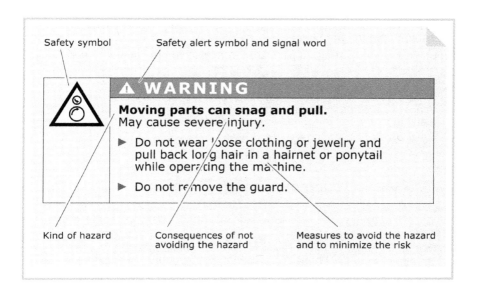

Safety symbol

Safety alert symbol and signal word

⚠ **WARNING**

Moving parts can snag and pull.
May cause severe injury.

▶ Do not wear loose clothing or jewelry and pull back long hair in a hairnet or ponytail while operating the machine.

▶ Do not remove the guard.

Kind of hazard

Consequences of not avoiding the hazard

Measures to avoid the hazard and to minimize the risk

Basic rules

- Always place the warning *directly before* the step that's dangerous or causes a danger. Users mostly follow instructions step by step. If the warning comes after a step, it may come too late. If the warning comes at the beginning of the topic, long before the instruction, readers might not read it at all or may have forgotten it when they come to the dangerous step.

- If a warning is independent of particular procedures and refers to working with the device in general, put it into a special safety instructions topic at the beginning of the document.

- Keep warnings short and to the point. The optimal length is one sentence. Avoid warnings that are longer than 3 sentences. Consider splitting longer warnings into 2 separate warnings. If explanations or instructions are needed, put them into other sections after the caution.

- When necessary, don't hesitate to repeat something that you have already said somewhere else.

- Invest much effort in writing clearly and simply (see *Writing sentences* 151) and *Writing words* 171).

- Invest much effort in writing so that readers who speak the document's language only as a second language can precisely understand the hazards.

- Address readers directly using the imperative form (see *Talk to the reader* 102). Don't use the passive voice (see *Use the active voice* 118).

- Use appropriate vocabulary to point out the possible consequences of disregarding the warning.

- Don't use contractions; in particular, don't use contractions with *not*. For example, use *do not* instead of *don't*, use *must not* instead of *mustn't*, and so on.

- Write a warning only if an action involves any real danger for people or things. To provide information that helps to prevent mistakes that don't result in injury, death, or damage, write a note instead.

- Don't use warnings in place of tips. Write a tip instead.

Types of warnings: CAUTION, WARNING, DANGER

The signal word at the beginning of the warning indicates the severity of the danger. To make it particularly emphatic, it's often written in capital letters.

- A warning that begins with the signal word **CAUTION** indicates a hazard that, if not avoided, *might* result in *minor* or *moderate* injury. A warning with the signal word **CAUTION** can also refer to a situation that could damage or destroy the product or the users' work. If a hazard doesn't involve any risk for people but only for things, the warning symbol is often omitted or the signal words **SAFETY INSTRUCTIONS** are used instead of **CAUTION** (ANSI Z535.6).

- A warning that begins with the signal word **WARNING** indicates a hazard that, if not avoided, *could* result in *death* or *serious injury*.

- A warning that begins with the signal word **DANGER** indicates a hazard that, if not avoided, *will* result in *death* or *serious injury*.

✔ Yes: ⚠ *CAUTION*
The lamp may get very hot.
To avoid skin burns, wait a few minutes before exchanging a defective lamp, or wear protective gloves.

⚠ *CAUTION*
Magnetic field.
May damage electronic components.
Keep electronic components away.

⚠ *CAUTION*
Formatting the disk deletes all data that have previously been stored on this disk.
Make a backup copy and store it in a safe place before you proceed.

✔ Yes: ⚠ *WARNING*
Ultraviolet radiation.
May cause irreversible eye damage.
Do not look directly at the UV lamp.

⚠ *WARNING*
Magnetic field.
Can be harmful to pacemaker wearers.
Maintain a distance of at least 30 cm (12 in.) from the
equipment.

✔ **Yes:**   ⚠ *DANGER*
High voltage.
Contact will cause electric shock, burns, or death.
Disconnect all power sources before removing the panel.

▬ (DE)

Im Deutschen lauten die entsprechenden Signalwörter:

Englisch	Deutsch
SAFETY INSTRUCTIONS	**Hinweis**
	alternativ:
	Achtung
	Sicherheitshinweise
CAUTION	**Vorsicht!**
WARNING	**Warnung!**
DANGER	**Gefahr!**

Hinweis:
Eine Schreibung in Großbuchstaben ist im Deutschen eher unüblich, stattdessen stehen – insbesondere bei Gefahr für Leib und Leben – häufig Ausrufezeichen.

When to use exclamation points

Usually, you don't need to use exclamation points within warnings because the warning symbol and the signal word already add enough emphasis.

However, if the warning symbol and the signal word are missing (not recommended), do add an exclamation point. You can also use an exclamation point if you want to put some special emphasis on a particular statement.

✔ **Yes:**   ⚠ *WARNING*
Hot surface.
Risk of burns.
Do not touch!

4.2.6 Writing cross-references and links

When writing cross-references and links, be aware of their advantages and disadvantages.

Advantages of cross-references and links are:

- they make it easy to find related information
- they make it possible to keep topics short; you can give optional or related information in another, linked topic

Disadvantages of cross-references and links are:

- they attract attention and interrupt the flow of reading
- they require a decision from readers whether or not to follow the link
- if readers follow the link, they often do so without having read the complete topic; if they don't return, they miss what comes after the link
- if readers don't follow the link, they might also miss some important information—or they might feel as though they are missing something

Readers are constantly faced with a dilemma: Should they yield to the alluring call of a link? Should they leave the current topic, or should they read on?

If readers decide to click a link but the topic that opens up doesn't contain the expected information, they feel that the document doesn't answer their questions and eventually stop reading.

Basic rules

The most important basic rules are:

- Use cross-references and links sparingly.
- Always link to a particular topic or page. Don't add references such as "... is described elsewhere" or "... can be found in another help topic."
- Choose a link text that's clear in its meaning even when a user doesn't read the surrounding text. Formulate link texts so that the content and information type of the target topic are clearly recognizable.
- Keep link texts simple and concise.
- When readers click a link, something new and unknown awaits them. Give them a sense of security by making sure that at least one keyword from the link text corresponds to the title of the subject invoked. This will affirm to them that they've landed in the right place, which creates confidence and

contributes to a positive user experience.

- Make links unobtrusive. When possible, integrate links seamlessly into the sentence so that the sentence doesn't change because of the link.

- Place links preferably at the end of sentences, paragraphs, and topics. In these places, they don't interrupt the readers.

- If you create printed manuals and online help from the same text base (single source publishing), avoid using media specific terms like *topic* or *chapter*.

- Embed only very important links into the running text. Move the majority of links out of the text into a separate navigation zone. This approach has a number of advantages:

  - readers' attention isn't diverted from the text that they're reading

  - formulation is significantly simpler, especially if you want to generate a printed manual and online help from the same text base (single source publishing)

  - administration is significantly simpler (this depends on the authoring tool)

  - translation is significantly simpler

  As a rule of thumb, include only those links in the main body of the text that really justify any interruption in the flow of the text. Ask yourself the following question: If you weren't writing but talking in a face-to-face conversation, would you interrupt yourself to draw the other person's attention to some additional information? If the answer is "yes," the link is probably justified in the text. If the answer is "no," move the link to a separate list of related topic links below the topic's body text.

Cross-references in printed manuals vs. links in online help

In a printed manual, a cross-reference needs a page number. Most authoring tools add the page number automatically when you generate the final document. To keep cross-references as short as possible, don't include any chapter numbers.

✘ No: ... *chapter 2.3.4 (page 75)*

✘ No: ... *in chapter 2.3.4 on page 75*

✔ Yes: ... *(page 75)*

✔ Yes: ... *on page 75*

If you want to generate a printed manual and online help from the same text base (single source publishing), also don't include any words that are typical for a printed manual or for online help in your source files, such as the word "page." Powerful single source authoring tools are capable of adding these terms automatically as required.

After production, the final results in your documents should look something like this:

✔ Yes: Online help: *You can find additional information about the graphics format under <u>The right format for your project</u>.*

✔ Yes: Printed manual: *You can find additional information about the graphics format under "The right format for your project" on page 225.*

✔ Yes: Online help: *Before you can use this function, you need an additional <u>license key</u>.*

✔ Yes: Printed manual: *Before you can use this function, you need an additional license key (see "Acquiring additional license keys" on page 13).*

Making link texts clear

Give readers some decision aid:

- Where will a link take them?
- What kind of information can they expect there?
- For which applications do they need this information?
- Will there be plenty of information, or just the bare minimum?
- Will the linked information be easy or difficult to understand?
- Will there be practical examples or abstract theory?

The more of these questions you can answer even before a reader clicks the link, the better.

Because many readers just scan online help text briefly, it's generally not enough if the information about the link is given in the accompanying text. Make it present in the link text itself.

✘ No: *Click <u>here</u> for more information.*

✘ No: *Click here for <u>more information</u>.*

✘ No: *Click <u>here</u> for the results of our usability study.*

✔ Yes: *Most users prefer links at the end of a topic. This is confirmed by the <u>results of our usability study on navigation (3 pages)</u>.*

     (Even if a reader doesn't read the complete sentence, this link text clearly communicates what the link target is about. This outweighs the disadvantage that this link text is longer than the link text in the previous examples.)

Use short phrases such as the popular "(<u>more ...</u>)" only if they're located near a clear title, for example, in a table or tabular listing.

✖ **No:** *Choose whether you want to write a letter, create a presentation, or plan a project (more ...).*

✔ **Yes:** *You can use the program*

- *to write letters (more ...)*
- *to create presentations (more ...)*
- *to plan projects (more ...)*

Mention any access restrictions immediately after the link but don't make this part of the link text.

✔ **Yes:** *You can also send reports by email (Professional Edition only).*

Making links unobtrusive

Every link draws the attention of the reader away from the actual content of the topic:

- The layout stands out prominently (underlining).
- Readers are anxious about missing important information by not following the link.

In addition, sentences often are longer and syntactically more complex if there's a link within them.

The ideal link is integrated into the text in such a way that the sentence would read just as if the link wasn't there. Try to write the link so that it only contains its content, but no extra navigational information.

✖ **No:** *To learn how to enter text, follow the link to the topic Using the On-Screen Keyboard.*

✔ **Yes:** *To enter text, use the on-screen keyboard.*

Don't advertise links. There's no need to explicitly refer to a link in the text—it stands out well enough on its own.

✖ **No:** *Click here to start the tutorial.*

✔ **Yes:** *You can find a step-by-step introduction in the Tutorial.*

Linking to pictures and tables

When possible, avoid linking to pictures and tables altogether. Try to place pictures and tables so that it's clear which picture or table the text relates to. In particular, avoid linking to pictures and tables that are in other topics.

If you can't avoid linking to a picture or table, usually the best solution is to add the plain link at the end of a sentence.

✘ No: *Insert the batteries as shown in figure 4 on page 23.*

✔ Yes: *Insert the batteries (fig. 4 on page 23).*

▬ (DE)

✘ Nein: *Legen Sie die Batterien, wie in Abbildung 4 auf Seite 23 gezeigt, ein.*

✔ Ja: *Legen Sie die Batterien ein (Abb. 4 auf Seite 23).*

4.3 Writing sentences

Form sentences that are easy to process for the human brain.

Rules for writing on the sentence level

- *Make short sentences* 152
- *Put the main thing into the main clause* 155
- *Avoid parentheses and nested sentences* 157
- *Feel free to repeat a word* 160
- *Add syntactic cues* 163
- *Be clear about what you're referring to* 165
- *Don't use "and/or"* 167

 (DE)

- *(DE) Vermeiden Sie „zerrissene Verben"* 169

4.3.1 Make short sentences

Follow the rule "One idea, one sentence."

Don't put too much information into one sentence.

Long sentences make it difficult to understand a text because they consume a lot of short-term memory.

It's not so much of a problem if you have just one long sentence. It's a big problem, however, if you have a succession of many long sentences.

Instructions (procedures) need especially short sentences because users must both read and act almost simultaneously, which limits their short-term memory. Never describe more than two actions within one sentence.

Write so that your reader has to read each sentence only once. Most readers aren't willing to read a sentence a second time.

- If readers who didn't understand a sentence *do* read a sentence a second time: They'll be angry that you've stolen their precious time.

- If readers who didn't understand a sentence *don't* read a sentence a second time: They'll miss an important piece of information. Maybe they even fail to use your product correctly.

Short sentences force you, the author, to be clear. This can be hard when writing a text, but it's essential for your readers. If you can't shorten all sentences, even occasional shorter sentences are helpful.

What's a good sentence length?

As a rule of thumb, 10 to 15 words per sentence is a good *average* sentence length.

In normal text, individual sentences with subordinate clauses may be up to about 25 words long.

In instructions, the longest sentences shouldn't be longer than approximately 20 words.

Tip:
When in doubt, read your own sentence. If you can't repeat the sentence from memory, it's too long.

There may be additional factors that further limit the acceptable maximum sentence length:

- Users with a low educational level need shorter sentences than users with a high educational level.

- Users who speak the document's language as a second language need shorter sentences than native speakers.
- Users who have little time to read a text need shorter sentences—especially in stressful and dangerous situations. (Example: Instructions on a fire extinguisher.)
- Users who are standing in a hot, noisy production hall need shorter sentences than users who are sitting in a quiet office or at home.

Tips for splitting long sentences

- Search for the words *and* and *or*. Often, these words signal the easiest and most sensible places to split a long sentence.
- Convert comma-separated lists into bulleted lists (see *Writing lists* 138) or tables.
- Convert "if-then" structures into bulleted lists.
- Convert sentences that describe multiple actions into a step-by-step procedure (see *Writing procedures* 128).

✖ **No:**   *You can use the program to write letters and you can use it to display and print presentation slides.*

✔ **Yes:**   **You can use the program to write letters. You can also use the program to display and print presentation slides.**

✔ **Yes:**   *You can use the program:*

- *to write letters*
- *to display presentation slides*
- *to print presentation slides*

✖ **No:**   *The manual is outdated, it was written 10 years ago by a trainee.*

✔ **Yes:**   **The manual is outdated. It was written 10 years ago by a trainee.**

✖ **No:**   *You have the option to print all pages, to print a range of pages, or to print the current page only.*

✔ **Yes:**   *You can print either:*

- *all pages*
- *a range of pages*
- *the current page only*

✖ **No:**   *Press button A if you want to perform action A, press button B if you want to perform action B, press button C if you want to perform action C, or press button D if you want to perform action D.*

✔ **Yes:** *Press one of the following buttons:*

To ...	press ...
(action A)	button A
(action B)	button B
(action C)	button C
(action D)	button D

✘ **No:** *To print the report, first open the report file, then choose **File** > **Print**, and finally click **OK**.*

✔ **Yes:** *To print the report:*

  1. Open the report file.

  *2. Choose **File** > **Print**.*

  *3. Click **OK**.*

4.3.2 Put the main thing into the main clause

Many writers alternate simple sentences with more complicated sentences that consist of a main clause plus one or more subordinate clauses. The advantage of this style is that then the text doesn't sound monotonous and boring. The disadvantage, however, is that you trade in variety for clarity.

Using sentences with subordinate clauses only makes sense in longer texts.

Don't use sentences with subordinate clauses at all:

- for the key information
- for things that are particularly difficult to understand

Use subordinate clauses only for those facts that are easy to understand and of minor importance.

Put the main fact into the main clause, not into a subordinate clause.

Position the key information at the beginning of the sentence.

✖ **No:**   *The PIN, which must not be noted on the credit card, consists of 4 digits.*

✔ **Yes:**   *The PIN consists of 4 digits. Don't note the PIN on the credit card.*

✖ **No:**   *Finally, you need to press the red button, which triggers the explosion.*

✔ **Yes:**   *To trigger the explosion, press the red button.*

▬▬ (DE)

Ersetzen Sie im Deutschen einfache Nebensätze nicht durch andere, kompliziertere Konstruktionen. Wenn es um einfache Zusammenhänge von Ursache und Wirkung geht, ist im Deutschen ein einfacher Nebensatz oft die bessere Lösung.

Geht es jedoch um komplexe Zusammenhänge, teilen Sie den Satz komplett auf.

✖ **Nein:** *Das zur Speicherung der Daten benutzte Dateiformat hängt von der Art der Datensätze ab.*

✔ **Ja:**   *Welches Dateiformat das Programm zum Speichern der Daten verwendet, hängt von der Art der Datensätze ab.*

✖ **Nein:** *Das zur Speicherung der Daten benutzte Dateiformat sowie das zur Übertragung genutzte Protokoll hängen von der Art der Datensätze ab.*

✔ **Ja:**   *Die Art der Datensätze bestimmt:*

- *welches Dateiformat das Programm zum Speichern der Daten verwendet*
- *welches Protokoll das Programm zum Übertragen der Daten verwendet*

✖ **Nein:** *Dies kann infolge des eingestellten Timeouts zu Fehlermeldungen beim Speichern von großen Dateien auf langsamen Speichermedien führen.*

✔ **Ja:** *Der eingestellte Timeout kann zu Fehlermeldungen führen, wenn Sie große Dateien auf langsamen Speichermedien speichern.*

4.3.3 Avoid parentheses and nested sentences

Parenthetic remarks and subordinate clauses nested into a main clause make it difficult to understand a sentence. Readers must:

- memorize the beginning of the sentence
- read and mentally process the parentheses
- recall the memorized beginning of the sentence and combine it with the rest of the sentence

1 Avoid parentheses and nested sentences, regardless of whether you put them inside commas, dashes, or parentheses. If the parenthetical remark is actually important, create a separate sentence. If the parenthetical remark isn't important, omit it.

2 Parentheses are OK for:

- introducing acronyms
- adding units of measure
- referring to numbered callouts in images

3 If you have a good reason to use a nested sentence, watch the punctuation. Sometimes there are commas within one sentence that have different roles. For example, you may have a parenthetical remark that includes a comma-separated list. When two separate functions are nested inside of each other, replace one set of commas with semicolons, parentheses, or dashes.

1

✖ **No:** *Many popular programs, for example, office suites, image editors, web browsers, and email clients, still come with poor documentation.*

✔ **Yes:** *Many popular programs still come with poor documentation. These include office suites, image editors, web browsers, and email clients.*

2

✔ **Yes:** *Prefer direct current (DC) if you want to ….*

✔ **Yes:** *Enter the size (in mm).*

✔ **Yes:** *Use the emergency switch (1) only if ….*

3

✖ No: *The names of user interface controls, such as menu items, tab cards, buttons, list boxes, etc., are printed in italics.*

(Here, the outer commas are used for the parentheses; the inner commas are used for the list of interface controls. This isn't wrong but makes the sentence difficult to read.)

✔ Yes: *The names of user interface controls (such as menu items, tab cards, buttons, and list boxes) are printed in italics.*

✔ Top: *The names of user interface controls are printed in italics. Typical user interface controls are menu items, tab cards, buttons, and list boxes.*

(DE)

Im Deutschen sind verschachtelte Nebensätze besonders problematisch, da hier das Verb oft erst am Ende des Satzes folgt.

Häufig funktioniert im Deutschen einer der folgenden Lösungsansätze:

- Trennen Sie den Satz in mehrere Sätze auf.
- Machen Sie aus einem eingeschobenen Nebensatz einen angehängten Nebensatz.
- Setzen Sie einen Doppelpunkt oder tragen Sie Umstandsangaben nach, zum Beispiel mit Formulierungen wie: *und zwar*, *das heißt*, *nämlich*, *besonders*.
- Stellen Sie das Verb möglichst nahe an den Satzanfang.

✖ Nein: *Bei extremer Kälte sollten vor dem Start zunächst Wischwasser, Kühlwasser sowie der Zustand der hinteren und vorderen Wischerblätter kontrolliert werden.*

✔ Ja: *Kontrollieren Sie bei extremer Kälte vor dem Start: Wischwasser, Kühlwasser, Zustand der vorderen und hinteren Wischerblätter.*

✖ Nein: *Zum Verputzen der Wand können Sie Raupputz, Kunstharzputz, oder sogar einfachen Kalk-Zementmörtel verwenden.*

✔ Ja: *Zum Verputzen der Wand können Sie Raupputz verwenden, Kunstharzputz oder sogar einfachen Kalk-Zementmörtel.*

✖ Nein: *Die Lager müssen unter Einhaltung der vorgeschriebenen Arbeitsreihenfolge mit Spezialfett, mindestens einmal jährlich oder alle 10000 km, geschmiert werden.*

✔ Ja: *Schmieren Sie die Lager mindestens einmal jährlich oder alle 10000 km mit Spezialfett. Halten Sie sich dabei an die*

---

vorgeschriebene Arbeitsreihenfolge.

✖ **Nein:** *Derjenige, der demjenigen, der diesen Satz, der in einem Kapitel, das in einem Buch, das schon vor Jahren besser nicht hätte geschrieben werden sollen, vorhanden ist, steht, geschrieben hat, besseres Deutsch beibringt, erhält eine Belohnung.*

✔ **Ja:** **Derjenige erhält eine Belohnung, der demjenigen besseres Deutsch beibringt, der ...**

Besonders kritisch sind im Deutschen auch restriktive Relativsätze. Anders als im Englischen fehlt hier eine Unterscheidung durch Kommasetzung und durch die gezielte Verwendung der Wörter *that* oder *which*. In der Folge sind solche Sätze oft mehrdeutig.

✖ **Nein:** *Fehler vom Typ „kritisch", die zum Programmabbruch führen, protokolliert das Programm in der Log-Datei.*

(Unklar: Protokolliert das Programm alle Fehler vom Typ „kritisch" oder nur solche, die zum Programmabbruch führen?)

✔ **Ja:** **Fehler vom Tpy „kritisch" führen zum Programmabbruch. Das Programm protokolliert sie in der Log-Datei.**

oder:

Solche Fehler vom Typ „kritisch", die zu einem Programmabbruch führen, protokolliert das Programm in der Log-Datei.

4.3.4   Feel free to repeat a word

When it adds clarity, don't hesitate to use the same word over again.

Don't avoid using the same word twice in a sentence or in consecutive sentences because you think that this is poor style. In user assistance, the best style is clarity.

1 In sentences with *and* and *or*, make sure that it's clear which terms an attribute relates to.

2 Words such as *this*, *that*, *they*, *these*, *those*, *it*, *which*, and so on link ideas. Make sure that these pointers point unmistakably to one noun, phrase, or clause so that your sentence isn't ambiguous. If there's *any* possibility of confusion, repeat the noun.

3 Don't use awkward constructions like *the former ... the latter*. These constructions aren't ambiguous, but they're extremely hard to read. Most readers do remember the statements, but they don't remember which one came first. So they have to read the whole section again, which wastes their time, makes them feel stupid, and doesn't contribute to a positive user experience.

1

✖ No: *You need green paper and tape.*

            (Unclear: Must the tape also be green?)

✔ Yes: *You need green paper and some tape.*

            or:

            You need green paper and green tape.

2

✖ No: *She told her colleague that her phone wasn't working.*

            (Ambiguous: Whose phone wasn't working? Her own phone or her colleague's phone?)

✔ Yes: *She told her colleague that the colleague's phone wasn't working.*

            or:

            She told her colleague that her own phone wasn't working.

✖ **No:** *You can use the instrument to measure the PH value and the humidity of the soil. Note that this only works if the temperature is above 0 degrees Celsius.*

(Unclear: Do both measurements need a minimum temperature? Also unclear: Which temperature is important? The temperature of the soil? The air temperature? Maybe even both temperatures?)

✔ **Yes:** *You can use the instrument to measure the PH value and the humidity of the soil. Note that you can only measure the PH value if the soil temperature is above 0 degrees Celsius.*

or:

You can use the instrument to measure the PH value and the humidity of the soil. Note that you can only measure the humidity if the soil temperature is above 0 degrees Celsius.

or:

You can use the instrument to measure the PH value and the humidity of the soil. Note that both measurements only work if the soil temperature is above 0 degrees Celsius.

3

✖ **No:** *Type 657B devices are green. Type 657C devices are red. The former are made of plastics, whereas the latter are made of steel.*

✔ **Yes:** *Type 657B devices are green. Type 657C devices are red. Type 657B devices are made of plastics, whereas type 657C devices are made of steel.*

✔ **Top:** *Type 657B devices are green and made of plastics. Type 657C devices are red and made of steel.*

(DE)

Verzichten Sie auch auf das Weglassen von Wortteilen und wiederholen Sie stattdessen diese Wortteile – bei Bedarf auch mehrfach. Zwar wird der Text dadurch geringfügig länger, die Wiederholung verbessert Lesbarkeit und Textverständnis jedoch deutlich.

✖ **Nein:** *Mit diesem Schalter können Sie das Gerät ein- und ausschalten.*

✔ **Ja:** *Mit diesem Schalter können Sie das Gerät einschalten und ausschalten.*

✔ **Top:** *Mit diesem Schalter schalten Sie das Gerät ein und aus.*

✖ **Nein:** *Das Gerät zunächst auf- und nach der Reinigung wieder zuschrauben.*

✔ **Ja:** *Schrauben Sie das Gerät auf, reinigen Sie es, und schrauben Sie es wieder zu.*

161

✔ **Ja:**   *1. Gerät aufschrauben.*
    2. Gerät reinigen.
    3. Gerät zuschrauben.

4.3.5 Add syntactic cues

Syntactic cues are words or punctuation marks that help readers to analyze the structure of a sentence more quickly and more reliably.

- Syntactic cues enhance readability.
- Syntactic cues can be especially helpful for readers who speak the document's language as a second language.
- Syntactic cues often eliminate ambiguities.

If you can add a syntactic cue, do so. The few extra characters or words are a good investment in clarity.

However, inserting a syntactic cue isn't always the best remedy for a poorly designed sentence. Sometimes it's better to rephrase a sentence completely or to make two sentences out of one (see *Make short sentences* 152).

The best way to get a good feeling for syntactic cues is to look at some examples:

✘ **No:** *Programs currently running are indicated by icons in the Task bar.*

✔ **Yes:** *Programs **that are** currently running are indicated by icons in the Task bar.*

✘ **No:** *The advanced search feature is especially helpful for users familiar with regular expressions.*

✔ **Yes:** *The advanced search feature is especially helpful for users **who are** familiar with regular expressions.*

✘ **No:** *You can print reports using the print utility.*

✔ **Yes:** *You can print reports **by** using the print utility.*

✘ **No:** *You can run macros using the Macro utility.*

(Ambiguous: Does the Macro utility run the macros or do the macros use the Macro utility?)

✔ **Yes:** *You can run macros **by** using the Macro utility.*

or

*You can run macros **that** use the Macro utility.*

✖ No: *The program continues processing data after restoring the database.*

✔ Yes: **The program continues to process data after it has restored the database.**

✖ No: *The operating system terminates the program if an error or exception occurs.*

✔ Yes: **The operating system terminates the program if an error or an exception occurs.**

✔ Top: **The operating system terminates the program if either an error or an exception occurs.**

✖ No: *If you choose option A and the program runs in auto-detection mode, something happens.*

✔ Yes: **If you choose option A and if the program runs in auto-detection mode, something happens.**

✖ No: *Your data must not include leading blanks and semicolons.*

(Ambiguous: Does the sentence mean only leading semicolons or all semicolons?)

✔ Yes: **Your data must not include semicolons and leading blanks.**

(Note: The syntactic cue here is the reversed order of *semicolons* and *blanks*.)

or:

Your data must not include leading blanks and leading semicolons.

✖ No: *The program was first published by company A and then modified by company B.*

✔ Yes: **The program was first published by company A and was then modified by company B.**

✖ No: *full- and part-time workers*

✔ Yes: **full-time and part-time workers**

✔ Top: **full-time workers and part-time workers**

4.3.6　Be clear about what you're referring to

When using words like *this*, *these*, *that*, *those*, *it*, *they*, and *them*, make sure that it's clear which subject you're referring to.

If it avoids ambiguity or improves readability: Don't hesitate to repeat the subject as often as necessary (see *Feel free to repeat a word* 160).

Keep your text as stupid as possible. Usually, the topics that you're talking about are challenging enough.

✖ **No:**　*The bomb is connected to a red and to a blue wire. Cut it to defuse the bomb.*

✔ **Yes:**　*The bomb is connected to a red and to a blue wire. To defuse the bomb, cut the red wire.*

or:

The bomb is connected to a red and to a blue wire. To defuse the bomb, cut the blue wire.

✖ **No:**　*DemoSoft can do A and B. This helps you to ....*

(Unclear: Does the word *this* relate to A, or does it relate to B, or does it relate to the combination of A and B?)

✔ **Yes:**　*DemoSoft can do A and B. A helps you to ....*

or

DemoSoft can do A and B. B helps you to ....

or

DemoSoft can do A and B. Both help you to ....

✖ **No:**　*The main problem that people run into with pronouns is not tying them to nouns.*

(Unclear: Who isn't tied to nouns: the people or the pronouns?)

✔ **Yes:**　*The main problem that people run into with pronouns is not tying the pronouns to nouns.*

✔ **Yes:**　*The main problem that a writer can run into with pronouns is not tying them to nouns.*

(Here you can use *them* because it can't relate to *writer*, which is singular.)

▬ (DE)

Auch im Deutschen existiert analog das Problem unklarer Bezüge, insbesondere im Zusammenhang mit den Wörtern *dies*, *diese*, *dieses*, *das*, *den*, *dem*, *es* und *sie*.

✘ Nein: *Studentin sucht Zimmer mit Bett, in dem sie auch Nachhilfe geben kann.*

(Unklar: Bezieht sich *in dem* auf das Zimmer oder auf das Bett?)

✔ Ja: *Studentin sucht Zimmer mit Bett. Das Zimmer soll auch für Nachhilfestunden geeignet sein.*

oder:

Studentin sucht Zimmer, in dessen Bett sie auch Nachhilfe geben kann.

Auch durch das Fehlen einer visuellen Information (Bild, Video, Vorführung) oder durch die in der Schriftsprache fehlende Möglichkeit ein Wort besonders zu betonen, kann ein Bezug unklar sein:

✘ Nein: *Trennen Sie das Kabel oben durch.*

✔ Ja: *Trennen Sie das Kabel an dessen oberen Ende durch.*

oder:

Trennen Sie das obere der vorhandenen Kabel durch.

Hinweis:
Als gesprochener Text zu einem Video wäre auch der ursprüngliche Text „Trennen Sie das Kabel oben durch" eindeutig. Allerdings sollten Sie auch in diesem Fall eine sprachlich eindeutige Version bevorzugen. Nicht jeder Betrachter schaut ohne Unterbrechung aufs Bild – insbesondere nicht bei handlungsanleitenden Videos. Außerdem verliert die Aussage Ihres Videos durch mehrdeutige Texte überflüssigerweise an Schärfe und Präzision, was sich negativ auf den Gesamteindruck auswirkt.

4.3.7 Don't use "and/or"

The phrase *and/or* is imprecise.

Either it's *and*, or it's *or*, but it can't be both. Don't burden the reader with the trouble of finding out what applies. It's your job to find out and then to communicate the facts clearly.

1 Replace the phrase *and/or* either with just *and* or with just *or*. If that's not possible, rephrase the sentence and explain the facts in more detail.

2 The same rules apply if *and/or* is implied in similar constructions with other words.

Note:
If you use *and/or*, there's no space character before and after the slash.

1

✖ **No:** *You can archive and/or delete reports.*

✖ **No:** *You can archive/delete reports.*

✔ **Yes:** **You can archive and delete reports.**

    or:

    You can archive or delete reports.

    or:

    You can first archive and then delete reports.

2

✖ **No:** *The procedure can read/write data.*

✔ **Yes:** **The procedure can read and write data.**

✖ **No:** *If the water is too hot/cold, adjust the temperature.*

✔ **Yes:** **If the water is too hot or too cold, adjust the temperature.**

✖ **No:** *If there's no input/output signal, ...*

✔ **Yes:** **If there's neither an input signal nor an output signal, ...**

    or:

    If there's either no input signal or no output signal, ...

(DE)

Im Deutschen gilt Analoges auch für das Wort *beziehungsweise* (siehe *Be specific* 109).

4.3.8   (DE) Vermeiden Sie „zerrissene Verben"

Formulieren Sie Sätze so, dass alle Bestandteile eines Verbs möglichst nahe beieinanderstehen.

Gleiches gilt für Verb plus Modalverben wie *dürfen*, *können*, *müssen*, *sollen*, *wollen*.

Dadurch, dass die Bestandteile der Verben nahe beieinanderstehen, entlasten Sie das Kurzzeitgedächtnis der Leser. Die Leser verstehen dann den Satzinhalt sofort – nicht erst am Satzende, nachdem alle Bestandteile des Verbs genannt wurden.

✖ **Nein:**   *Fügen Sie alle Objekte, wie z. B. Linien, Rechtecke, Dreiecke und Kreise, die Sie übernehmen möchten, in Ihr Bild ein.*

(In diesem Satz wurde *einfügen* zerrissen.)

✔ **Ja:**   *Fügen Sie alle Objekte in Ihr Bild ein, die Sie übernehmen möchten, z. B. Linien, Rechtecke, Dreiecke und Kreise.*

(Zwar stehen auch hier die Verbteile (*fügen ... ein*) nicht unmittelbar nebeneinander, sind jedoch zumindest näher zusammengerückt und stehen zudem im selben Teilsatz.)

✖ **Nein:**   *Dies kann infolge des eingestellten Timeouts zu Fehlermeldungen beim Speichern von großen Dateien auf langsamen Speichermedien führen.*

(In diesem Satz wurde *führen können* zerrissen.)

✔ **Ja:**   *Der eingestellte Timeout kann zu Fehlermeldungen führen, wenn Sie große Dateien auf langsamen Speichermedien speichern.*

✖ **Nein:**   *Das Fahrzeug kann, wenn Treibstoff in den Motorraum gelangt, explodieren.*

(In diesem Satz wurde *explodieren können* zerrissen.)

✔ **Ja:**   *Das Fahrzeug kann explodieren, falls Treibstoff in den Motorraum gelangt.*

✖ **Nein:**   *Sie dürfen diesen Satz, wenn Sie dazu in der Lage sind, gerne verbessern.*

(In diesem Satz wurde *verbessern dürfen* zerrissen.)

✔ **Ja:**   *Sie dürfen diesen Satz gerne verbessern, wenn Sie dazu in der Lage sind.*

4.4   Writing words

Use words that are common, simple, and unambiguous.

Rules for writing on the word level

- *Use short, common words* 172
- *Watch for "…ed"* 174
- *Watch for "the … of" and for "of the"* 175
- *Watch for opening "It …" and "There …"* 176
- *Avoid abbreviations and acronyms* 177
- *Use technical terms carefully* 181
- *Always use the same terms* 182
- *Avoid strings of nouns* 184
- *Avoid stacks of modifiers* 185
- *Avoid unnecessary qualification* 186
- *Use strong verbs* 188
- *Use fair language* 192

 (DE)

- *(DE) Umgang mit Anglizismen* 195

4.4.1   Use short, common words

Don't show off your vocabulary. Reading infrequently used words requires more mental work than reading common words. This slows down readers and makes your text hard to understand.

Keep it simple and stupid (the KISS principle).

Use words that are common and short. Use uncommon and long words only if there is no simpler alternative.

Bear in mind readers who speak the document's language only as a second language.

Typical examples of words that you should replace with simpler alternatives are nouns with 3 or more syllables and foreign words.

Instead of the complex term ...	use the simple form ...
alphabetical character	*letter*
attempt	*try*
capability	*ability*
condition	*state*
determination	*choice*
employ	*use*
indicate	*show*, *tell*, *say*
indication	*sign*
inspect	*check*
location	*site*, *place*
modify	*change*
preserve	*keep*
require	*need*
terminate	*end*
transmit	*send*
utilization	*use*

(DE)

Typische Beispiele in deutscher Sprache sind insbesondere durch Vor- und Zusätze aufgeblähte Wörter sowie Fremdwörter.

Anstelle des komplexen Ausdrucks ...	verwenden Sie besser die einfache Form ...
abändern	ändern
absinken	sinken
abklären	klären
ansteigen	steigen
aufzeigen	zeigen
aufspalten	spalten, teilen
Aufgabenstellung	Aufgabe
die allermeisten	die meisten
Fraktur	Bruch
frühzeitig	früh
Rücksichtnahme	Rücksicht
Unkosten	Kosten
Zielsetzung	Ziel

4.4.2 Watch for "...ed"

Try to simplify or omit words that end with "...ed."

If "...ed" indicates passive voice, put the sentence into active voice (see *Use the active voice* 118).

✘ No: *Users who are located in France ....*
✔ Yes: *Users in France ....*

✘ No: *A study that was conducted by our company shows that ....*
✔ Yes: *A study by our company shows that ....*

✘ No: *centralized control*
✔ Yes: *central control*

✘ No: *improved results*
✔ Yes: *better results*

4.4.3 Watch for "the … of" and for "of the"

> When possible, avoid the words *of* and *the*.
>
> Exception: Possessives from company names, product names, feature names, and objects. However, in this case it's often better to use the name of the company, product, feature, or object as an adjective.

✖ **No:**   *The purpose of this program is to print reports.*
✔ **Yes:**   *This program prints reports.*

✖ **No:**   *The assembly of computers is often done in China.*
✔ **Yes:**   *Computers are often assembled in China.*

✖ **No:**   *Some of the tasks require special knowledge.*
✔ **Yes:**   *Some tasks require special knowledge.*

✖ **No:**   *DemoSoft's key features*
✔ **Yes:**   *the key features of DemoSoft*

✖ **No:**   *Enter the disk's name.*
✔ **Yes:**   *Enter the name of the disk.*
✔ **Top:**   *Enter the disk name.*

4.4.4   Watch for opening "It …" and "There …"

> Review sentences that start with:
>
> - "It is …."
> - "There is …."
> - "There are …."
>
> Often, you can find a more concise solution.

✘ No: *It is often the case that sentences are much too long.*
✔ Yes: ***Often, sentences are too long.***

✘ No: *There are some functions that help you to save energy.*
✔ Yes: ***Some functions help you to save energy.***

✘ No: *There's something wrong with this sentence.*
✔ Yes: ***Something is wrong with this sentence.***

 (DE)

Im Deutschen existiert das Problem ganz analog.

Zusätzlich versteckt sich im Deutschen ein einleitendes *es* häufig auch hinter anderen Satzanfängen.

✘ Nein: *Es ist ratsam, diesen Satz besser zu formulieren.*
✔ Ja: ***Formulieren Sie diesen Satz besser.***

✘ Nein: *Konstruktionsbedingt kann es vorkommen, dass sich hinter der Abdeckung Schmutzpartikel anlagern.*
✔ Ja: ***Konstruktionsbedingt können sich hinter der Abdeckung Schmutzpartikel anlagern.***

4.4.5 Avoid abbreviations and acronyms

When in doubt, spell it out.

The advantage of improved clarity far outweighs the disadvantage of longer text. The use of too many abbreviations and acronyms is one of the most frequent reasons why readers fail to understand manuals and dislike reading them.

Abbreviate terms only if they appear in narrow table cells or other tight spaces.

Use acronyms only:

- if the acronym is also used in other materials that you can't change
- if the acronym is also used in the user interface of your product
- if you're sure that all readers know the meaning of the acronym
- if you need to use the term that the acronym stands for *very* often

If you can't avoid using an acronym

When you use an acronym, spell it out the first time, and then add the acronym in parentheses.

- In a printed manual, the "first time" is the page with the lowest page number.
- In online help, the "first time" is the topic that will be used either the earliest or the most frequently.

When spelling out the acronym, don't capitalize the words that make up the acronym unless the spelled-out form is a proper noun.

✔ **Yes:**   *original equipment manufacturer (OEM)*

✔ **Yes:**   *World Wide Web Consortium (W3C)*

If the pronunciation of an acronym isn't evident, provide a hint.

✔ **Yes:**   *WYSIWYG (pronounced "wiz-zee-wig")*

✔ **Yes:**   *W3C (W three C)*

Use capital letters without periods (exception: some geographic names).

✔ **Yes:**   *USB*

✔ **Yes:**   *XML*

✔ **Yes:** *EU*

✔ **Yes:** *U.S.*

✔ **Yes:** *U.K.*

To form the plural of an acronym, add a lowercase *s* without an apostrophe.

✔ **Yes:** *PCs*

✔ **Yes:** *CPUs*

Use an apostrophe only if you need to form a possessive of an acronym.

✔ **Yes:** *the OEM's products*

Don't include a generic term after an acronym if one of the acronym's letters stands for the same term.

✖ **No:** *HTML language*

    (Note: the letter *L* already stands for the word *language* because HTML is the acronym for *Hypertext Markup Language*.)

✔ **Yes:** *HTML*

✔ **Yes:** *Hypertext Markup Language (HTML)*

▬ (DE)

Beachten Sie im Deutschen, dass Sie auch für englische Akronyme einen bereits im Akronym enthaltenen Begriff nicht wiederholen.

✖ **Nein:** *PIN-Nummer*

    (Der Buchstabe *N* steht hier bereits für *Nummer*.)

✔ **Ja:** *PIN*

✖ **Nein:** *FAQ-Fragen*

    (Der Buchstabe *Q* steht hier bereits für *Questions = Fragen*.)

✔ **Ja:** *FAQ*

Handling common abbreviations

In English, there are a number of common Latin abbreviations for frequently used word and phrases, such as "i.e." for "that is" or "e.g." for "for example."

Even though these Latin abbreviations are very common, avoid them when possible. Many readers don't know which words these terms abbreviate or they confuse the abbreviations. In particular, "i.e." and "e.g." are often confused.

Usually, the few extra characters needed to spell out the words are a good investment in clarity. Use Latin abbreviations only in situations where there isn't enough space to spell out the words, such as in narrow table columns.

Instead of ...	use ...
i.e.	*that is*, *in other words*
e.g.	*for example*
etc.	*and so on* Note: It's OK to use *etc.* for most audiences because it can't be confused with any other abbreviation.
et al.	*and others*
cf.	*compare*
viz.	*namely*
vs., v.	*versus*, *as opposed to* Note: It's acceptable to use *vs.* in headings.

 (DE)

Da im Deutschen keine lateinischen Abkürzungen üblich sind, können Sie hier etablierte Abkürzungen bedenkenlos verwenden – nicht nur bei Platzmangel, sondern auch im normalen Fließtext.

Hinweis:
Bei mehrteiligen Abkürzungen stehen Leerzeichen zwischen den Bestandteilen. Verwenden Sie hierzu geschützte Leerzeichen, bei denen kein automatischer Zeilenumbruch erfolgt. Bei der Verwendung einer nichtproportionalen Schrift (z. B. Courier) kann das Leerzeichen entfallen, bei Verwendung einer Proportionalschrift gilt ein fehlendes Leerzeichen in gedruckten Texten jedoch gemeinhin als falsch. Falls möglich, können Sie auch ein etwas schmaleres Leerzeichen als das normale Leerzeichen verwenden.

Schreiben Sie folgende Wendungen *nicht* aus:

- nicht *zum Beispiel*, sondern *z. B.*
- nicht *das heißt*, sondern *d. h.*
- nicht *unter anderem*, sondern *u. a.*
- nicht *und so weiter*, nicht *etc.*, sondern *usw.*

Schreiben Sie folgende Wendungen *nur dann aus*, wenn Sie nicht zusammen mit einem konkreten Zahlenwert auftreten:

- *circa*, aber *ca. 10 cm*
- *minimal*, aber *min. 900 €*
- *maximal*, aber *max. 160 km/h*

Schreiben Sie alle anderen Abkürzungen *im Fließtext immer aus*, z. B.: *im Allgemeinen*, *in der Regel*, *insbesondere*, *ausschließlich*, *einschließlich*, *positiv*, *negativ*, *gegebenenfalls*, *sogenanntes*, *siehe oben*, *siehe unten*.

4.4.6   Use technical terms carefully

Technical terms speed up communication between people who share the same expertise. For others, the same technical terms are just incomprehensible.

Use technical terms only:

- if you're writing for experts
- if a term is also used in other materials for the same user group as your document and you can't change these materials
- if a term is used in the user interface of your product

If you can't avoid using technical terms, explain them when using them for the first time.

- In a printed manual, the "first time" is the page with the lowest page number.
- In online help, the "first time" is the topic that will be used either the earliest or the most frequently.

✖ No: *This opens your default email client.*

(You don't need the technical term *client* here.)

✔ Yes: *This opens your default email program.*

✔ Yes: *Plug the cable into the USB port of your computer.*

(Here it's OK to use the technical term *USB port* because it's the only common name that exists for this type of interface. You can't change this name. It doesn't make any sense to invent a new name.)

4.4.7   Always use the same terms

If you mean the same thing, use the same term.

If you use different terms, readers may think that you mean different things. Consistent terminology is also important when users want to skim your text for particular information, and when they use full-text search.

Variety is excellent for novels and for English tests at school. In technical writing, however, there's little room for creativity. Keep it simple and stupid (the KISS principle).

If you're writing sales texts and want to make your texts more conversational and vivid, vary the unimportant words rather than the important words.

Make a terminology list of the terms that you use and of the terms that you don't use (see *Be consistent* 111).

Don't use the terms that are on your blacklist within the visible text of your document, but add them as index keywords so that they appear in the alphabetical index. Readers who don't yet know which particular term you use can then find a topic even if they're looking for the "wrong" term.

Decision aids

When in doubt:

- use the word that's used in the user interface or printed on the product
- use the word that your users will be looking for when they skim a text
- use the simpler word
- use the shorter word

✖ **No:**   *If you've purchased a new application, you must first install the software before you can use the program.*

✔ **Yes:**   *If you've purchased a new program, you must first install this program before you can use it.*

Examples

Some typical examples of terms that need a decision on which words you use are:

- Do you say *computer*, or *PC*, or *machine*, or *client*, or *workstation*, or *unit*?

- Do you say *sound adapter*, or *sound card*?

- Do you say *pointer*, or *mouse pointer*, or *cursor*, or *mouse cursor*, or *arrow*, or *mouse arrow*, or *I-beam*?

- Do you say *keyboard shortcut*, or *quick access key*, or *shortcut key*, or *accelerator key*, or *hotkey*, or *speed key*, or *fast key*, or *quick key*, or *key combination*?

 (DE)

Typische Beispiele aus dem Deutschen:

- *Handbremse* oder *Feststellbremse*?

- *Blinker* oder *Fahrtrichtungsanzeiger*?

- *Produkt* oder *Artikel*?

- *Produktpreis* oder *Artikelpreis* oder *Produktkosten* oder *Artikelkosten*, oder …?

- *Button* oder *Schalter* oder *Schaltfläche*?

4.4.8 Avoid strings of nouns

> Strings of nouns are hard to understand and sometimes even ambiguous.
>
> Only use strings of nouns when they're the names of systems and you don't have the authority to simplify these names.
>
> When a string of nouns is used as an adjective, use hyphens for clarification.

✖ **No:** *the device adapter card port signals*

(This is ambiguous. We can't tell whether it refers to "signals of the port" or to "port signals of the card.")

✔ **Yes:** *the device-adapter-card port signals*

✔ **Top:** *the port signals of the device adapter card*

✖ **No:** *technical documentation writing principles*

✔ **Yes:** *principles of writing technical documentation*

✔ **Yes:** *Local Area Network*

(You can't change this technical term.)

 (DE)

Im Deutschen sind aus mehreren Wörtern zusammengesetzte Substantive häufig noch problematischer als im Englischen. Unübersichtlich wird es insbesondere dann, wenn keine Bindestriche zur Verbesserung der Lesbarkeit eingefügt wurden (siehe auch *(DE) Bindestriche*).

✖ **Nein:** *Festplattencontrolleranschlussklemme*

✔ **Ja:** *Anschlussklemme des Festplatten-Controllers*

(Hinweis: Der komplexe zusammengesetzte Ausdruck wurde hier in zwei einfach zu bewältigende Wörter aufgetrennt. Zusätzlich wurden, zur besseren Lesbarkeit, in *Festplattencontroller* die deutsche und die englische Komponente durch einen (optionalen) Bindestrich getrennt.)

4.4.9 Avoid stacks of modifiers

When two or more modifiers appear before a noun, the meaning of the phrase often becomes ambiguous.

Using a hyphen in the right place can sometimes resolve the ambiguity. However, the phrase remains difficult to read and hard to understand.

For this reason: If there are two or more modifiers before a noun, always try to rephrase the sentence.

✘ **No:**   *Typical data conversion problem areas include ....*

✔ **Yes:**   *Typical data-conversion problem areas include ....*

✔ **Top:**   *Areas in which users typically have problems when converting data include ....*

✘ **No:**   *more effective methods*

✔ **Yes:**   *methods that are more effective*

or:

more methods that are effective

4.4.10   Avoid unnecessary qualification

Don't modify or qualify words that don't need to be modified or qualified.

Unnecessary qualification only adds empty calories, but it doesn't add any valuable information.

✖ **No:**   *Both sentences tell you exactly the very same thing.*

✔ **Yes:**   *Both sentences tell you the same.*

Additional examples:

Instead of ...	use the simple form ...
absolutely unique	*unique*
completely identical	*identical*
completely new	*new*
entirely complete	*complete*
exactly alike	*alike*
highly innovative	*innovative*
integral part	*part*
lift up	*lift*
new innovation	*innovation*
perfectly clear	*clear*
precisely the same	*the same*
repeat again	*repeat*
round circle	*circle*
the reason why	*the reason*
totally new	*new*
visible to the eye	*visible*

(DE)

Typische Beispiele in deutscher Sprache:

Statt …	verwenden Sie besser nur …
aktive Mitarbeit	Mitarbeit
andere Alternativen	Alternativen
die allermeisten	die meisten
Eigeninitiative	Initiative
exemplarisches Beispiel	Beispiel
frühzeitig	früh
gemachte Erfahrungen	Erfahrungen
gestellte Aufgabe	Aufgabe
hundertprozentig sichergehen	sichergehen
innerer Kern	Kern
nähere Einzelheiten	Einzelheiten
neu renoviert	renoviert
nochmals wiederholen	wiederholen
restlos überzeugt	überzeugt
Rückantwort	Antwort
Rückerstattung	Erstattung
sich zurück erinnern	sich erinnern
wie z. B.	wie
Zukunftsprognose	Prognose

4.4.11 Use strong verbs

> Use strong verbs that keep your texts clear, simple, and concise.
>
> Many people tend to use "smothered verbs" because they feel that these verbs make their text sound more sophisticated. A smothered verb is a verb that is converted into a noun, which is then made the object of a less precise verb. (See the examples below.)
>
> Avoid all sorts of smothered verbs.

✘ No: *We held a meeting and reached a decision on the improvement of our documents.*

✔ Yes: *We met and decided on how to improve our documents.*

✘ No: *The first step is the deletion of all needless words.*

✔ Yes: *The first step is to delete all needless words.*

✘ No: *Our software can be of help to you.*

✔ Yes: *Our software can help you.*

✘ No: *There are four screens within the wizard.*

✔ Yes: *The wizard consists of four screens.*

✘ No: *You can exert influence on the printing quality by using different types of paper.*

✔ Yes: *You can influence printing quality by using different types of paper.*

Additional examples:

Instead of ...	use the stronger, simpler verb ...
achieve reductions	*reduce*
conduct an analysis	*analyze*
conduct an investigation of	*investigate*
do an inspection of	*inspect, check*
form a plan	*plan*
give an answer to	*answer*
have knowledge of	*know*

have reservations about	*doubt*
have a concern	*worry*
hold a meeting	*meet*
make a decision	*decide*
make a distinction	*distinguish*
make a proposal	*propose*
make a recommendation	*recommend*
make a suggestion	*suggest*
provide a solution	*solve*
reach an agreement	*agree*

 (DE)

Im Deutschen entstehen ähnliche Probleme durch **Nominalisierungen**. Vermeiden Sie insbesondere:

- Sätze mit **Verben** wie *erfolgen, geschehen, passieren, sich ereignen*
- **Substantive** mit der Endung *-ung*

✘ Nein: *Zum Schutz Ihrer Daten erfolgt eine tägliche Datensicherung.*

   (unklar: Wer handelt?)

✔ Ja: **Zum Schutz Ihrer Daten sichert das Programm die Daten automatisch einmal täglich.**

   oder:

   Sichern Sie Ihre Daten täglich, um einem Datenverlust vorzubeugen.

✘ Nein: *Das Legen von Eiern erfolgt durch Hühner, während das Geben von Milch durch Kühe vorgenommen wird.*

✔ Ja: **Hühner legen Eier. Kühe geben Milch.**

✘ Nein: *Das Gerät dient der Überwachung des Blutdrucks sowie des Herzschlags.*

✔ Ja: **Das Gerät überwacht Herzschlag und Blutdruck.**

✘ Nein: *Alle 3 Monate ist eine Reinigung des Luftfilters erforderlich.*

✘ Nein: *Alle 3 Monate ist der Luftfilter zu reinigen.*

   (Keine gute Alternative, da Passiv.)

✖ Nein: *Alle 3 Monate muss der Luftfilter gereinigt werden.*

(Keine gute Alternative, da Passiv.)

✔ Ja: *Reinigen Sie den Luftfilter alle 3 Monate.*

alternativ Telegrammstil:

Luftfilter alle 3 Monate reinigen.

✖ Nein: *Aufgrund des Preisanstiegs im Rohstoffbereich mussten auch die Preise für unsere Produkte angepasst werden.*

✔ Ja: *Da die Rohstoffpreise gestiegen sind, mussten wir auch die Preise unserer Produkte anpassen.*

✖ Nein: *Die Nichtbefolgung dieser Regeln kann zu schweren Verletzungen führen.*

✔ Ja: *Wenn Sie diese Regeln nicht befolgen, können Sie sich oder andere Personen schwer verletzen.*

✖ Nein: Überschrift: *Verwaltung von Gebäuden*

✔ Ja: Überschrift: *Verwalten von Gebäuden*

(*Verwalten* ist hier zwar auch substantivisch gebraucht, ist jedoch schon näher am Verb als *Verwaltung*.)

✔ Top: Überschrift: *Gebäude verwalten*

(echtes, starkes Verb)

Weitere Beispiele:

Statt ...	verwenden Sie besser die einfache Form ...
einer Prüfung unterziehen	*prüfen*
Einfluss nehmen	*beeinflussen*
Gebrauch machen von	*benutzen*
in der Lage sein	*können*
in Erwägung ziehen	*erwägen*
in Frage stellen	*bezweifeln*
ist abhängig von	*hängt ab von*
ist ausreichend	*reicht aus*
ist störend	*stört*
Messung durchführen	*messen*
Möglichkeit eröffnen	*ermöglichen*

Problemlösung herbeiführen	*Problem lösen*
Schutz bieten vor	*schützen vor*
Sie haben die Möglichkeit	*Sie können*
zum Abschluss bringen	*abschließen*

4.4.12   Use fair language

Be aware of the variety of people who use your product.

- Write in a gender-neutral way.
- Avoid cultural biases and stereotypes relating to religion, family structure, leisure activities, purchasing power, and a particular lifestyle.
- In examples, use both female and male first names. Use last names that reflect different cultural backgrounds.

Avoid using phrases like *he or she* or *his/her* in your attempt to be gender neutral. Phrases like these attract even more attention to gender and thus partly defeat your purpose. (Also, they bloat the text and slow readers down.)

To be gender neutral, the following often helps:

- Use a gender-neutral term instead of a gender-specific one.
- Don't speak of *users*. Address your readers directly as *you*.
- Put a sentence into the plural.

Note:
When using the plural to refer to a specific group of people, do so consistently throughout the whole document even if there's no problem with the gender in a particular case (see also *Be consistent* 111). Another advantage of the plural form is that it's often shorter because you don't need *a* or *the*.

✘ **No:**   *If there's no sound in your earphones, ask the stewardess for assistance.*

✔ **Yes:**   **If there's no sound in your earphones, ask the flight attendant for assistance.**

✘ **No:**   *Each user must enter his password.*
✘ **No:**   *Each user must enter her password.*
✘ **No:**   *Each user must enter his/her password.*
✘ **No:**   *Each user must enter their password.*
✔ **Yes:**   **Each user must enter his or her password.**
✔ **Top:**   **All users must enter their passwords.**

or:

Users must enter their passwords.

✖ **No:** *When a user wants to print a report, he or she needs to choose the Print command from the File menu.*

✔ **Yes:** *When users want to print a report, they need to choose the Print command from the File menu.*

✔ **Top:** *If you want to print a report, choose File > Print.*

or:

To print a report, choose File > Print.

Additional examples of gender-neutral terms:

Instead of ...	use the gender-neutral form ...
actress	*actor*
businessman	*businessperson, professional*
chairman	*chair, chairperson*
delegates and their wives	*delegates and their spouses*
fireman	*firefighter*
fisherman	*fisher*
foreman	*supervisor*
handyman	*caretaker, repairer*
hostess	*host*
housewife	*homemaker*
mailman	*letter carrier*
man hours	*working hours, work hours*
man-made	*synthetic, artificial, hand-made, manufactured*
manpower	*workers, workforce, staff, employees, personnel*
policeman	*police officer*
repairman	*repairer, technician*
salesman, salesgirl, saleslady	*salesperson, representative*
stewardess	*flight attendant*
tradesman	*tradesperson*
waitress	*waiter, server*
watchman	*security guard*
workmen	*workers*

193

(DE)

Der Trick, einen Ausdruck einfach in die Mehrzahl zu setzen, damit er geschlechtsneutral wird, funktioniert im Deutschen meist nicht.

Auch ist es im Deutschen oft deutlich schwieriger als im Englischen, eine geschlechtsneutrale Bezeichnung anstelle einer geschlechtsspezifischen Bezeichnung zu finden (Beispiele: *Studierende* statt *Studentinnen und Studenten*, *Lesende* statt *Leserinnen und Leser*).

Vermeiden Sie unbedingt schwer lesbare oder avantgardistische Konstrukte wie *Benutzer/-innen* oder *BenutzerInnen*. Auch durch eine durchgehend weibliche Form („als Monteurin wissen Sie ...") ist nichts gewonnen, denn hiervon fühlen sich wiederum männliche Leser irritiert oder amüsiert.

Wenn Sie tatsächlich geschlechtsneutral schreiben wollen, besteht die einzig brauchbare Lösung in einer Konstruktion mit *und* und beiden Geschlechtsformen („Monteurinnen und Monteure"). Dies ist wegen der gesteigerten Textlänge aber nur dann zu empfehlen, wenn es selten vorkommt. Frei von Problemfällen ist allerdings auch diese Lösung nicht: Was machen Sie z. B. mit einem Wort wie *Arbeitnehmervertreter?* Schreiben Sie dann *Arbeitnehmervertreter und Arbeitnehmervertreterinnen* – oder *Arbeitnehmervertreter und Arbeitnehmerinnenvertreter* – oder *Arbeitnehmervertreter und Arbeitnehmerinnenvertreterinnen* ...?

Am einfachsten und damit am benutzerfreundlichsten werden Ihre Dokumente, wenn Sie durchgängig die männliche Form im neutralen Sinne verwenden (Beispiel: *Benutzer*). Im Vorwort oder im Impressum können Sie darauf z. B. mit folgendem Text hinweisen:

»In diesem Text verwenden wir die männliche Form in einem neutralen Sinne und verzichten auf Doppelformen wie in „Leser/-innen". Wir bitten alle Leserinnen um Verständnis für diese Vereinfachung.«

Auch für Sie als Autor ist diese Lösung im Deutschen am einfachsten.

4.4.13   (DE) Umgang mit Anglizismen

Anglizismen sind aus dem Englischen übernommene Wörter und Redewendungen.

Viele Leser empfinden Anglizismen als befremdlich oder verstehen Sie nur ungenau.

Verwenden Sie Anglizismen wie alle Fremdwörter nur dann, wenn es dafür einen wichtigen Grund gibt.

Verwenden Sie Anglizismen *nicht*:

- um Kompetenz, Agilität oder Modernität zu suggerieren
- um sich vor einer klaren Aussage zu drücken

Verwenden Sie Anglizismen *bewusst*:

- wenn es keine geläufige deutsche Entsprechung gibt, wenn also ein Wort mit deutschen Wörtern nur langatmig oder unvollkommen umschrieben werden kann
- wenn das englische Wort bei Ihrer Zielgruppe bereits geläufiger ist als das entsprechende deutsche Wort
- wenn es sich um eine feststehende Produktbezeichnung oder Teilebezeichnung handelt, die Sie selbst nicht beeinflussen können

Typische Gruppen von Anglizismen

Zu den Anglizismen gehören:

- Übernommene Wörter
  Beispiele: *Software*, *Browser*, *scannen*
- Übernommene Redewendungen
  Beispiele: *Sinn machen* („to make sense") statt *sinnvoll sein*, *einmal mehr* („once more") statt *noch einmal*, *nicht wirklich* („not really") statt *eigentlich nicht*.
- Pseudo-Anglizismen, die zwar englisch klingen, im Englischen jedoch eine andere Bedeutung haben oder gar nicht existieren.
  Beispiele: *Handy*, *Beamer*, *Public-Viewing*.

Groß oder klein?

Schreiben Sie englische Substantive in einem deutschen Text groß (einzige Ausnahme: innerhalb von Zitaten).

✔ **Ja:**   *Sie müssen dieses Tool verwenden, wenn Sie ...*

Schreiben Sie bei Zusammensetzungen mit Bindestrich aus
Substantiv+Substantiv beide Wortteile groß.

✔ **Ja:**   *Tool-Abhängigkeit* (alternativ: *Toolabhängigkeit*)

Schreiben Sie bei Zusammensetzungen mit Bindestrich aus
Adjektiv+Substantiv den substantivischen Teil groß, den adjektivischen Teil
klein.

✘ **Nein:**   *tool-abhängig*

✔ **Ja:**   *Tool-abhängig* (alternativ: *toolabhängig*)

Schreiben Sie bei Getrenntschreibung aus Adjektiv+Substantiv beide Teile
groß.

✔ **Ja:**   *Compact Disc*

Zusammen oder getrennt? Mit oder ohne Bindestrich?

Prinzip Zusammenschreibung

Anders als im Englischen werden im Deutschen zusammengesetzte Substantive
prinzipiell zusammengeschrieben. Eine Getrenntschreibung ohne Bindestrich
ist im Deutschen falsch. Eine Zusammenschreibung ist grundsätzlich immer
richtig, allerdings oft schlecht lesbar.

✘ **Nein:**   *Web Site*

✔ **Ja:**   *Website*

✘ **Nein:**   *Software Installation*

✔ **Ja:**   *Softwareinstallation*

Optionale Bindestriche zur Verbesserung der Lesbarkeit

Um die Lesbarkeit zu verbessern, *können* Sie zwischen Wortgliedern
Bindestriche einfügen. Dies gilt sowohl für rein englisch-englische
Zusammensetzungen als auch für gemischt englisch-deutsche oder deutsch-
englische Zusammensetzungen.

✔ **Ja:**   *Software-Installation*

✔ **Ja:**   *Drag-and-drop-Unterstützung*

Sonderfälle

Bei Fügungen aus Abkürzung (oder Akronym) und Hauptwort *müssen* Sie einen
Bindestrich einfügen.

✔ **Ja:**   *USB-Kabel*

✔ **Ja:**   *EDV-lastig*

Ebenso *müssen* Sie einen Bindestrich einfügen bei Aneinanderreihungen von 3 oder mehr Bestandteilen und bei Kombinationen mit Präpositionen.

✔ **Ja:**   *Burn-out*

✔ **Ja:**   *Burn-out-Syndrom*

✔ **Ja:**   *Shut-down*

✔ **Ja:**   *Shut-down-Prozedur*

Verwenden Sie *keinen* Bindestrich bei Wörtern, die bereits im Englischen zusammengeschrieben werden.

✔ **Ja:**   *Handheld*

✔ **Ja:**   *Touchscreen*

Wörter, die mit einem Adjektiv beginnen, können Sie optional auch komplett getrennt schreiben. Dies ist oft besonders übersichtlich.

✔ **Ja:**   *Compact Disc* (alternativ: *Compact-Disc*, *Compactdisc*)

✔ **Ja:**   *High Tech* (alternativ: *High-Tech*, *Hightech*)

✔ **Ja:**   *Black Box* (alternativ: Black-*Box*, Blackbox)

✔ **Ja:**   *Direct Banking* (alternativ: Direct-*Banking*, Directbanking)

Ziehen Sie mehrere Bestandteile nicht zu einem Wort mit zwei Großbuchstaben zusammen. Einzige Ausnahme: feststehende Produktnamen.

✖ **Nein:**   *CompactDisc*

✖ **Nein:**   *HighTech*

✔ **Ja:**   Produktname: *DemoSoft*

Männlich, weiblich oder sächlich?

Kriterium 1: Wortendung
Allgemein richtet sich das Geschlecht bei Fremdwörtern primär nach der Wortendung. Da Anglizismen allerdings häufig andere Endungen haben als die klassischen Fremdwörter, ist diese Methode nicht immer eindeutig.

Kriterium 2: Geschlecht der deutschen Übersetzung
Ist eine Zuordnung des Geschlechts über die Wortendung nicht möglich, richtet sich das Geschlecht nach der deutschen Übersetzung. Beispiel: *das Pattern*, denn die entsprechende Übersetzung ist „das Muster". Allerdings können auch hier Zweifelsfälle entstehen, nämlich dann, wenn es zu einem Wort Übersetzungen mit unterschiedlichem Geschlecht gibt. Beispiel: *die E-Mail* oder *das E-Mail* (Übersetzungen: „die Post" oder „das Schreiben").

Kriterium 3: Häufigkeit der Verwendung
Falls auch das Geschlecht der deutschen Übersetzung nicht eindeutig ist,

müssen Sie, falls nötig, in einem Wörterbuch nachschlagen. Finden Sie dort den Begriff nicht, suchen Sie in einer Internet-Suchmaschine nach beiden infrage kommenden Alternativen. Richten Sie sich im Zweifelsfall nach der Variante mit der größeren Anzahl an Treffern.

Kriterium 4: Im Zweifel sächlich

Wenn alle zuvor genannten Kriterien kein befriedigendes Ergebnis liefern, bevorzugen Sie im Zweifelsfalle die sächliche Variante.

Einer Festlegung ausweichen

Wenn ein Anglizismus nur selten in Ihrem Dokument vorkommt, können Sie einer Festlegung oft auch ausweichen:

- Stellen Sie einen Absatz so um, dass der Anglizismus in der Mehrzahl erscheint.
  Beispiel: *Ihr/Ihre E-Mail versenden* wird zu ***E-Mails versenden***.

- Stellen Sie einen Absatz so um, dass Sie einen unbestimmten Artikel verwenden können (funktioniert nur bei Zweifeln zwischen männlich und sächlich).
  Beispiel: *der/das Cluster* wird zu ***ein Cluster***.

Wie die Mehrzahl bilden?

Die Mehrzahl fast aller Anglizismen bilden Sie durch Anhängen von *s*.

Dabei gelten *nicht* die Regeln der englischen Rechtschreibung, sondern die Regeln der *deutschen* Rechtschreibung, denn Sie schreiben einen *deutschen* Text. Ein *y* wird also nicht zu *ie*.

Die Mehrzahlbildung bei bereits in den deutschen Wortschatz übergegangenen Wörtern kann individuell abweichen. Hier hilft gegebenenfalls nur ein Blick in ein *deutsches* Wörterbuch.

✔ **Ja:**   *die Backups*
✔ **Ja:**   *die Hardwares*

✖ **Nein:** *die Queries*
✔ **Ja:**   *die Querys*

✖ **Nein:** *die Computers*
✔ **Ja:**   *die Computer*

Wie beugen?

Wenn Sie einen deutschen Text schreiben, beugen Sie auch englisch-stämmige Wörter grundsätzlich entsprechend der *deutschen* Rechtschreibung und Grammatik.

Verben:

✔ **Ja:**   ich *recycle*, du *recyclest*, wir *recyclen*

✔ **Ja:**   ich *designe*, du *designst*, wir *designen*

✔ **Ja:**   ich *mailte*, ich habe *gemailt*, ich werde *mailen*

Substantive:

✘ **Nein:**   *die Verwaltung des Network*

✔ **Ja:**   **die Verwaltung des Networks**

Substantivische Abkürzungen in der Einzahl können Sie auch ohne Genitiv-s schreiben. In der Mehrzahl müssen Sie aber immer ein s anfügen.

✔ **Ja:**   **das Kabel des PC(s)**

✔ **Ja:**   **die Preise der PCs**

✔ **Ja:**   **die Verschlüsselung des WLAN(s)**

✔ **Ja:**   **die Verschlüsselung der meisten WLANs**

Adjektive:

Viele Adjektive können nicht gebeugt werden. In diesen Fällen müssen Sie zwingend eine Ersetzung vornehmen.

✘ **Nein:**   *Die Arbeit wurde outgesourced.*

✔ **Ja:**   **Die Arbeit wurde outgesourct.**

✔ **Top:**   **Die Arbeit wurde ausgelagert.**

✘ **Nein:**   *ein pinkes Gehäuse*

✔ **Ja:**   **ein pinkfarbenes Gehäuse**

Allgemeine Tipps:

Generell ist es meist besser, einen Anglizismus ganz zu ersetzen, als eine eingedeutschte Beugung zu erzwingen, die zu einem kaum vertretbaren Ergebnis führt.

✘ **Nein:**   *Wenn Sie die Datei downgeloaded und die Software upgedated haben, ...*

✘ **Nein:**   *Wenn Sie die Datei gedownloaded und die Software geupdated haben, ...*

✔ **Ja:**   **Wenn Sie die Datei heruntergeladen und das Programm aktualisiert haben, ...**

✘ **Nein:**   *Ich habe Ihnen die E-Mail geforwarded.*

✔ **Ja:**   **Ich habe die E-Mail an Sie weitergeleitet.**

Wie trennen?

Trennen Sie nur Zitate und längere englische Textabschnitte entsprechend den englischen Regeln.

Trennen Sie ansonsten sprechsilbenorientiert.

Trennen Sie keine feststehenden Klangeinheiten.

✔ **Ja:**   *Brain|stor|ming*

✔ **Ja:**   *Down|loa|den*

✔ **Ja:**   *Mouse*

4.5   FAQ: Grammar and word choice

Even small errors affect your credibility and undermine the quality of your product.

Don't underestimate the importance of correct grammar and word choice. Flawless instructions are much more trustworthy than those with many errors.

In everyday speech, many words are used interchangeably. However, when you write instructions and technical documentation, choosing one word as opposed to another can make a vital difference.

> **ⓘ Important:** Use electronic grammar checkers with care. They can identify many mistakes, but they can't find them all. Electronic spelling checkers, grammar checkers, and other language tools don't work as a substitute for editing and proofreading by a human.

FAQ

When writing a text, you probably don't want to waste your time browsing bulky grammar reference manuals and textbooks. For this reason, we've compiled quick answers to the most frequent questions that arise when writing instructions and technical documents.

Tip:
The listed topics include not only frequently asked questions but also frequently made mistakes. Even if you don't have any particular question now, take the time to skim the topics in this section for details that you might not be aware of.

The terms are sorted alphabetically. If one topic covers several terms, the more common term is listed before the more uncommon term. If you're looking for a specific word and don't find it, look for it in the index.

4.5.1   accurate / precise

> Use *accurate* when you mean *correct, free from error*, or *true*.
>
> Use *precise* when you mean *minutely exact down to a couple of decimal points* or *sharply defined*. *Precise* also refers to conformance to a strict standard or pattern. Think of *precise* as a very narrow range of tolerance.

✔ **Yes:**   *This instrument is very accurate. It always shows the correct value.*

✔ **Yes:**   *This instrument is precise. It shows the value with a precision of 10 decimals.*

✔ **Yes:**   *You can't be too accurate, but you can be too precise if you exceed the number of significant figures in your calculations.*

4.5.2   allow / enable

People *allow* things, but things *enable* people.

Use *allow* when you mean *permit* or *consent to*. Also use *allow* when you mean *to allocate a certain amount*.

Use *enable* when you mean *provide the means or power*, or *make something possible or easy*. In this case, it's often better to use *can* or *let*.

✔ **Yes:**   *You're allowed to make one backup copy of the program.*

✔ **Yes:**   *The trial version of the program allows you to create only documents that have 3 pages or less.*

✔ **Yes:**   *Leaving some space in the drum allows for expansion of the liquid when the liquid is hot.*

✘ **No:**   *This program allows you to hack your competitor's web site.*

✔ **Yes:**   *This program enables you to hack your competitor's web site.*

✔ **Top:**   *With this program, you can hack your competitor's web site.*

4.5.3 and / as well as / plus

Usually, use *and*.

However, *as well as* and *plus* can make it easier for readers to predict and analyze the structure of the rest of a sentence.

Use *as well as* to set off different items in a list.

Avoid *as well as* when it might be confused with *as good as*, especially if you're writing for an international audience.

✖ No: *You can use the machine to cut wood as well as steel.*
✖ No: *You can use the machine to cut wood plus steel.*
✔ Yes: *You can use the machine to cut wood and steel.*
✔ Yes: *You can use the machine to cut wood, steel, and plastics.*

✔ Yes: *We sell computers, software, and T-shirts.*
✔ Yes: *We sell computers and software, as well as T-shirts.*

(Note: There are two different classes of goods: technical products and clothing.)

✔ Yes: *The form lets you enter your name and other personal data.*

(Note: After the word *and*, this sentence might also continue very differently. Example: "... and if you aren't careful you might end up on a spammer's list.")

✔ Top: *The form lets you enter your name, as well as other personal data.*

(Note: Here, after *as well as*, the reader can expect a noun. This makes the sentence more predictable and easier to read.)

✔ Yes: *The kit includes two cables and a selection of spare parts.*
✔ Top: *The kit includes two cables plus a selection of spare parts.*

(Note: The word *plus* clearly signals that you're talking about a list of things. If you use *and*, the sentence might also continue very differently. Example: "... and is one of the best on the market.")

✔ Yes: *You can write manuals as well as online help files.*

(Note: This sentence is ambiguous. Does it mean that you can write both manuals *and* help files? Or does it mean that you're good at writing manuals and equally good at writing online help files?)

✔ Top: *You can write manuals plus online help files.*

4.5.4   because / since / as

Use *because* to refer to a reason.

Use *since* only to refer to the passage of time. This avoids ambiguities and prevents misinterpretation.

Don't use *as* as a synonym for *because*. The word *as* can be ambiguous and has so many meanings that it's problematic, especially for an international audience.

✖ **No:**   *Since the paper is missing, you can't print.*

✖ **No:**   *As the paper is missing, you can't print.*

✔ **Yes:**   **Because the paper is missing, you can't print.**

✔ **Yes:**   **Since Monday, the paper has been missing.**

✔ **Yes:**   **Because we installed the program, we've saved a lot of time.**

(means: "We've saved a lot of time. The reason for this is the installed program.")

✔ **Yes:**   **Since we installed the program, we've saved a lot of time.**

(means: "We've saved a lot of time. This happened after we installed the program.")

4.5.5 can / may / might / must / should

Always be as precise as possible. The incorrect or vague use of the terms *can*, *may*, *might*, *must*, *should*, *could*, and so on is one of the most frequent causes for misinterpretation.

1 Use *can* when you mean the ability or power to do something.

2 *May* and *might* both indicate possibility or probability. *Might* suggests a somewhat lower probability than *may*.

Don't use *may* to imply the ability to do something. In this case, use *can*.

Don't use *may* to imply the permission to do something. In this case, use *allowed to*.

Tip:
Phrases with *you* include *can* more often than *may* ("You can ....").

3 Use *must* to describe a user action that's required. If you feel that the word *must* is too strong or impolite because it implies an obligation, rephrase your sentence as an instruction or use *need to* or *have to*.

Note:
In American English, don't use *shouldn't* instead of *mustn't* because you think that *mustn't* sounds too British. Use *must not* instead, which is unambiguous for an international audience.

4 Use *should* only to describe a user action that's recommended but optional. However, try to avoid the word *should* altogether because it always conveys an element of doubt. Instead, clearly tell the reader what to do, or clearly mark your sentence as a recommendation.

Never use the word *shall* in technical documentation.

> (DE)
>
> Achten Sie besonders bei der Übernahme oder Übersetzung von Texten aus dem amerikanischen Sprachraum darauf, dass im amerikanischen Englisch *shouldn't* oft gleichbedeutend mit *mustn't* (*must not* = „nicht dürfen") verwendet wird.

✖ No: *You may use a spoon or a fork.*

✔ Yes: *You can use a spoon or a fork.*

✘ **No:**   *You may use the program to write a manual.*
✔ **Yes:**   **You can use the program to write a manual.**

2

✔ **Yes:**   *You may be right.*
✔ **Yes:**   **Print quality may be poor if you use cheap paper.**
✔ **Yes:**   **The camera is shock proof, but it might break if you drop it from a height of more than 1.5 meters.**

✘ **No:**   *The remote control may also use a frequency of 140 MHz.*

(This is Ambiguous: It could mean that (a) you're allowed to use a frequency of 140 MHz; (b) the device is able to send at 140 MHz if you set it up to do so; (c) it might happen that the device sends at 140 MHz at its own discretion.)

✔ **Yes:**   **It's also allowed to use a frequency of 140 MHz for the remote control.**
(If you mean that law permits you to use this frequency.)

The remote control can also use a frequency of 140 MHz.
(If you mean that the device is able to send at this frequency.)

The remote control might also use a frequency of 140 MHz.
(If you mean that the device chooses the frequency automatically.)

✘ **No:**   *You may print the report, or you may save it to a file.*
✔ **Yes:**   **You can print the report, or you can save it to a file.**

3

✔ **Yes:**   **The fluid must pass the valve.**

✔ **Yes:**   **If the function doesn't work, you must contact support.**
✔ **Top:**   **If the function doesn't work, you need to contact support.**
✔ **Top:**   **If the function doesn't work, contact support.**

4

✘ **No:**   *You should make a backup copy of the file.*
✔ **Yes:**   **We recommend that you make a backup copy of the file.**
✔ **Top:**   *Make a backup copy of the file.*

✖ **No:** *You should get a reply within 24 hours.*

✔ **Yes:** **You will usually get a reply within 24 hours.**

✔ **Top:** **We usually reply within 24 hours.**

---

must / must not

For many writers who speak English as a second language, the use of *must* and *must not* is a dangerous pitfall. In some languages, the literal translation of *must not* means *need not*. In English, *must not* and its contraction *mustn't* mean that there's an **obligation not to do something**.

✔ **Yes:** **You must not smoke while filling up the tank.**

✔ **Yes:** **You must not use a 110-volt device on a 220-volt outlet.**

---

cannot / can not

Use *cannot* or its contraction *can't* if you mean *is not able to*. Here, the word *not* negates the word *can*.

Use *can not* if the word *not* relates to the action following the word *can*, not to the word *can*.

✔ **Yes:** **You cannot / can't start the compressor while the engine is running.**

(means: It isn't possible to start the compressor while the engine is running. It won't work.)

✔ **Yes:** **You can not start the compressor while the engine is running.**

(means: You don't have to start the compressor while the engine is running. It's not required, although it may be more common or recommended.)

---

can / could

The word *could* conveys an element of doubt, which is something that you must avoid in user assistance (see *Be specific* 105). For this reason, don't use *could* when you mean *can*, *does*, or *will*.

Only use *could* as the past tense of *can*. In most cases, however, it's better to use the present tense (see *Use the present tense* 117).

✖ **No:** *If you follow this rule, this could improve your documents.*

✔ **Yes:** **If you follow this rule, this improves your documents.**

✖ **No:** *Instead of using the menu, you could also use a keyboard shortcut to format the text.*

✔ **Yes:** *Instead of using the menu, you can also use a keyboard shortcut to format the text.*

✔ **Yes:** *The program couldn't find your name in its database.*

✔ **Top:** *The program can't find your name in its database.*

4.5.6 check / control

Use *check* when you mean *to examine* or *to make sure that something is correct, safe, or suitable*.

Use *control* when you mean *to direct, to operate*, or *to limit something*.

✔ **Yes:** *You can check the brightness of the display by pressing the Info key.*

(means: You can read a number there, but you can't change the brightness.)

✔ **Yes:** *You can control the brightness of the display by pressing the Settings key.*

(means: You can change the brightness there.)

✘ **No:** *Control whether all wires are connected correctly.*

✔ **Yes:** *Check whether all wires are connected correctly.*

✔ **Yes:** *The slider allows you to control the voltage.*

4.5.7   effective / efficient

Use *effective* when you mean *producing a desired effect*.

Use *efficient* when you mean *productive* or *with little waste*.

✔ **Yes:**   *Cooling a desktop computer with liquid helium is effective, but it isn't efficient.*

(Helium cools very well, so it's effective. However, helium is more expensive than other ways of cooling a computer, so helium isn't efficient.)

✔ **Yes:**   *An efficient power supply has a higher effective power output than an inefficient power supply.*

4.5.8   if / when / whether / whether or not

1 Use *if* to express a condition.

Use *when* for situations that require preparation. Also use *when* to denote the passage of time.

Use *whether* to express uncertainty and alternatives. Don't use *whether or not* in this case.

Only use *whether or not* if you want to emphasize that there are two possibilities or if you mean *under any circumstances*. However, it's often better to rephrase the sentence in this case.

2 Don't use *then* in if-clauses. It's a needless filler word and can even be misinterpreted in the sense that readers think that there's a time-based relation.

1

✔ **Yes:**   *If the engine starts ....*

   (means: It's not clear whether the engine will start.)

✔ **Yes:**   *When the engine starts ....*

   (means: At that point in time when the engine starts.

✔ **Yes:**   *If you need help, read the manual.*
✔ **Yes:**   *When Setup is complete, restart your computer.*
✔ **Yes:**   *To find out whether you need a new ink cartridge, ...*

✔ **Yes:**   *You must pay your taxes whether you want to or not.*
✔ **Top:**   *You must pay your taxes even if you don't want to.*

2

✘ **No:**   *If you need help, then read the manual.*
✔ **Yes:**   *If you need help, read the manual.*

4.5.9 like / such as / as

In everyday speech, *like* and *such as* are often used interchangeably. In technical documentation, however, be precise:

- *like* means *similar to*
- *such as* means *as for example*

Don't use *like* as a conjunction; use *as* instead.

✔ **Yes:** *Office suites like LibreOffice provide a powerful word processor.*

(means: Only those office suites provide a powerful word processor that are similar to LibreOffice. There may be other office suites that are different and that don't provide a powerful word processor.)

✔ **Yes:** *Office suites, such as LibreOffice, provide a powerful word processor.*

(means: All office suites provide a powerful word processor. LibreOffice is an example of an office suite.)

✔ **Yes:** *Moving a dialog box is like moving a window.*

✖ **No:** *You can work with remote files like you would with local files.*

✔ **Yes:** *You can work with remote files as you would with local files.*

4.5.10 safety / security

Use *safety* when referring to the protection from physical damage that's caused by natural disasters or accidents.

Use *security* when referring to the protection from damage that's caused by humans (often intentionally). This can be physical and non-physical damage.

✔ **Yes:** *Online banking is perfectly safe, but it's not always secure.*

✔ **Yes:** *job safety*

(Refers to the risk of being injured at the work place.)

✔ **Yes:** *job security*

(Refers to the risk of becoming unemployed.)

✔ **Yes:** *safety measures*

(Means, for example, the obligation to wear a helmet and protective goggles.)

✔ **Yes:** *security measures*

(Means, for example, measures against burglary and industrial espionage.)

4.5.11   that / which

> Don't use *that* and *which* interchangeably or to avoid word repetitions. Word repetitions are perfectly OK in technical writing (see *Always use the same terms* 182).
>
> - Use *that* for restrictive clauses (clauses that cannot be removed without distorting the meaning).
> - Use *which* for nonrestrictive clauses (clauses that can be put in parentheses or removed entirely).
>
> Put a comma before *which* but don't put a comma before *that*.

✔ **Yes:**   *Press the key that's labeled Ignition.*

✔ **Yes:**   *Press the Ignition key, which launches the rocket.*

✘ **No:**   *Delete the line that defines the command which you want to remove.*

✔ **Yes:**   *Delete the line that defines the command that you want to remove.*

Note that *that* or *which* can imply a very different meaning:

✔ **Yes:**   *We accepted the last bid, which was sent by fax.*

(means: We accepted the last bid. By the way, this bid was sent by fax.)

✔ **Yes:**   *We accepted the last bid that was sent by fax.*

(means: We accepted the last bid of those bids that were sent by fax. However, there might have been other bids that we received later by mail or by email.)

✔ **Yes:**   *The file, which stores the passwords, was deleted.*

(means: The file was deleted. Incidentally, this file stores the passwords.)

✔ **Yes:**   *The file that stores the passwords was deleted.*

(means: A particular file was deleted. It was the password file.)

4.5.12   use / utilize / employ

You *use* items when you use them the way they were intended to be used.

You *utilize* items when you use them in a way that they weren't designed for.

Employ suggests the use of a person or thing that was previously idle or inactive. To avoid confusion, especially with readers who speak English as a second language, don't use *employ* as a synonym for *use* or *utilize*. Use *employ* only when you mean *to hire*.

✘ No: *Utilize a screwdriver to tighten the screw.*

✔ Yes: *Use a screwdriver to tighten the screw.*

✔ Yes: *Utilize a screwdriver as a chisel.*

✘ No: *Employ a colleague to help you.*

✔ Yes: *Ask a colleague to help you.*

✔ Yes: *The boss employed two additional experts.*

4.5.13 (DE) Benutzer / Anwender

Durch die Eindeutschung des englischen Begriffs *Application* (*Anwendung*) als Synonym für *Programm* hat sich im Deutschen auch der Begriff *Anwender* stark verbreitet.

Bevorzugen Sie trotzdem den Begriff *Benutzer*.

Tipp:
In den meisten Fällen benötigen Sie weder den Begriff *Anwender* noch den Begriff *Benutzer*. Adressieren Sie Ihre Leser stets direkt mit *Sie*. Sprechen Sie nur dann von *Benutzern*, wenn Sie ein Dokument für Entwickler schreiben und in Ihrem Text die Personen meinen, für die die Entwickler ein Programm erstellen oder Gerät konstruieren (siehe auch *Talk to the reader* 102.)

✖ **Nein:** *Mit dieser Anwendung arbeiten mehr als 1000 Anwender.*

✔ **Ja:** *Mit diesem Programm arbeiten mehr als 1000 Benutzer.*

4.5.14 (DE) der / welcher

---

1 Verzichten Sie auf die Verwendung von *welcher*, *welche*, *welches* als Relativpronomen und verwenden Sie stattdessen immer *der* (*den*), *die* oder *das*.

Verwenden Sie *welcher*, *welche*, *welches* auch dann nicht als Ersatz für *der* (*den*), *die* oder *das*, wenn dies zu einer Doppelung führt (z. B. „... der, der ...").

Tipp:
Häufig können Sie eine Doppelung auch vermeiden, indem Sie das Substantiv wiederholen, was gleichzeitig die Lesbarkeit verbessert.

2 Verwenden Sie *welcher*, *welche*, *welches* nur, wenn es um eine Auswahl geht.

1

✖ **Nein:**   *Dies ist die Antwort, auf welche Sie so lange gewartet haben.*

✔ **Ja:**   *Dies ist die Antwort, auf die Sie so lange gewartet haben.*

✖ **Nein:**   *Aber ist es auch die, welche Sie sich wünschen?*

✔ **Ja:**   *Aber ist es auch die, die Sie sich wünschen?*

✔ **Top:**   *Aber ist es auch die Antwort, die Sie sich wünschen?*

2

✔ **Ja:**   *Entscheiden Sie, welche Datei Sie löschen möchten.*

4.5.15   (DE) durch / von

Das Wort *durch* können Sie häufig durch *von* oder eine treffendere Formulierung ersetzen.

✖ **Nein:**   *Die Maschine wird durch Menschen bedient.*
✔ **Ja:**   *Die Maschine wird von Menschen bedient.*

✖ **Nein:**   *Sie können das Gerät durch unseren Service reparieren lassen.*
✔ **Ja:**   *Sie können das Gerät von unserem Service reparieren lassen.*
✔ **Top:**   *Gerne kümmert sich unser Service um die Reparatur.*

4.5.16 (DE) mehr als / über / oberhalb

Um das Verstehen schon während des Lesens eines Satzes zu erleichtern:

- Verwenden Sie beim Bezug auf Mengen oder Größen grundsätzlich *mehr als*.

- Verwenden Sie *über* nur, wenn Sie sich auf eine räumliche Position direkt senkrecht zu einem Bezugspunkt beziehen.

- Verwenden Sie *oberhalb*, wenn Sie sich auf eine allgemein höher gelegene räumliche Position beziehen.

Hinweis:
Analoges gilt für *weniger als / unter / unterhalb*.

✖ **Nein:** *Das Gerät kann über 200 Bilder speichern.*

✔ **Ja:** *Das Gerät kann mehr als 200 Bilder speichern.*

✔ **Ja:** *Der Schalter befindet sich über der Buchse.*

  (bedeutet: Der Schalter befindet sich senkrecht darüber.)

✔ **Ja:** *Das Kabel befindet sich oberhalb der Abdeckung.*

  (bedeutet: Das Kabel befindet sich irgendwo auf der Abdeckung oder darüber.)

4.5.17 (DE) so

Der Vorteil des Wortes *so* liegt in seiner Kürze, der Nachteil in der großen Vielzahl möglicher Bedeutungen.

1 Ersetzen Sie das Wort *so* wenn möglich durch einen präziseren Ausdruck. Verwenden Sie das Wort *so* insbesondere nicht als Kurzform für *ebenso*.

2 Vermeiden Sie das Wort *so* am Anfang eines neuen Absatzes, denn es bezieht sich immer auf den vorangehenden Satz. Ein Absatz, der mit dem Wort *so* beginnt, ist daher beim Querlesen nicht verständlich.

3 Innerhalb eines Absatzes ist es meist unproblematisch, wenn ein Satz mit dem Wort *so* beginnt. In vielen Fällen können Sie damit umständlichere Formulierungen vermeiden und Ihren Text kurz halten.

4 Verwenden Sie das Wort *so* grundsätzlich nicht als Füllwort.

1

✘ Nein: *Die Bildqualität dieses Fernsehers ist so gut wie die Bildqualität eines teureren Modells.*

✔ Ja: *Die Bildqualität dieses Fernsehers ist ebenso gut wie die Bildqualität eines teureren Modells.*

2

✘ Nein: *So kann es passieren, dass das Gerät beschädigt wird.*

✔ Ja: *Wenn Sie die Batterien falsch herum einsetzen, kann dies das Gerät beschädigen.*

3

✘ Nein: *Mit der Funktion Mehrseitendruck können Sie mehrere Berichtsblätter auf einer Seite zusammenfassen. Auf diese Weise ist es möglich, eine Menge Papier zu sparen.*

✔ Ja: *Mit der Funktion Mehrseitendruck können Sie mehrere Berichtsblätter auf einer Seite zusammenfassen. So sparen Sie viel Papier.*

4

✘ Nein: *Wenn Sie den Bericht drucken wollen, so wählen Sie den Menüpunkt ...*

✔ **Ja:**   *Wenn Sie den Bericht drucken wollen, wählen Sie den Menüpunkt ...*

4.5.18 (DE) und / sowie

Prinzipiell können Sie *sowie* immer durch *und* ersetzen.

Das Wort *sowie* kann jedoch hilfreich sein, um unterschiedliche Gruppen in einer Liste klar voneinander abzutrennen.

✖ **Nein:** *Mit diesem Mobiltelefon können Sie Spiele spielen, Musik hören und telefonieren.*

(Hier werden die beiden Funktionsgruppen „Unterhaltung" und „Telefonie" nicht klar voneinander abgegrenzt.)

✔ **Ja:** *Mit diesem Mobiltelefon können Sie Spiele spielen und Musik hören sowie telefonieren.*

✖ **Nein:** *Wir verkaufen Computer, Software und T-Shirts.*

(Hier werden die beiden unterschiedlichen Warengruppen „Technik" und „Technik" nicht klar voneinander abgegrenzt.)

✔ **Ja:** *Wir verkaufen Computer, Software sowie T-Shirts.*

✔ **Ja:** *Wir verkaufen Computer und Software sowie T-Shirts.*

4.5.19 (DE) während / wohingegen

Um Zweifelsfälle zu vermeiden und einen Satz schon beim ersten Lesen verständlich zu machen, verwenden Sie *während* nicht als Synonym für *wohingegen*.

- Verwenden Sie das Wort *während* nur dann, wenn Sie einen zeitlichen Bezug meinen.

- Verwenden Sie *wohingegen*, wenn Sie einen Gegensatz ausdrücken möchten.

✖ **Nein:** *Modell A ist grün, während Modell B rot ist.*

✔ **Ja:** *Modell A ist grün, wohingegen Modell B rot ist.*

✔ **Ja:** *Sie können den Hebel nur bewegen, während Sie die Entriegelung gedrückt halten.*

✔ **Ja:** *Während das Programm die Daten überträgt, können Sie bedenkenlos weiterarbeiten.*

4.5.20 (DE) wenn / falls / sofern / sobald

1 Verwenden Sie für Bedingungen ausschließlich Bedingungssätze mit *wenn* oder *falls*. Dies verbessert die Lesbarkeit, denn das Verhältnis Bedingung-Folge wird sofort ersichtlich.

2 Verwenden Sie *falls*, wenn es ausschließlich um einen ursächlichen Zusammenhang geht

Verwenden Sie *wenn*, wenn bei der Bedingungen auch eine zeitliche Komponente eine Bedeutung hat.

Vermeiden Sie das gehobene *sofern*.

Verwenden Sie *sobald* nicht für Bedingungen, sondern nur dann, wenn Sie ein zeitliches Verhältnis ausdrücken möchten.

3 Verzichten Sie beim Nennen der Bedingung auf das Wort *dann*. Dies ist nicht nur ein überflüssiges Füllwort, sondern kann auch zu Missverständnissen führen, wenn es als Hinweis auf einen zeitlichen Zusammenhang fehlgedeutet wird.

1 **2**

✖ **Nein:** *Beim Erreichen der Arbeitstemperatur erlischt die Warnleuchte.*

✖ **Nein:** *Ist die Arbeitstemperatur erreicht, erlischt die Warnleuchte.*

✔ **Ja:** *Falls die Arbeitstemperatur bereits erreicht ist, erlischt die Warnleuchte.*

✔ **Ja:** *Wenn die Arbeitstemperatur erreicht ist, erlischt die Warnleuchte.*

✔ **Ja:** *Die Warnleuchte erlischt erst, wenn die Arbeitstemperatur erreicht ist.*

✔ **Ja:** *Die Warnleuchte erlischt, sobald die Arbeitstemperatur erreicht ist.*

✔ **Ja:** *Falls ich heute Abend nach Hause komme, bringe ich eine Pizza mit.*

(bedeutet: Ich weiß noch nicht genau, ob ich heute Abend nach Hause komme, aber falls ja, bringe ich eine Pizza mit.)

✔ **Ja:** *Wenn ich heute Abend nach Hause komme, bringe ich eine Pizza mit.*

(bedeutet: Ich komme heute Abend ganz sicher nach Hause. Wenn es soweit ist, bringe ich eine Pizza mit.)

3

✖ **Nein:** *Wenn Sie Hilfe brauchen, dann lesen Sie das Handbuch.*

✔ **Ja:** ***Wenn Sie Hilfe brauchen, lesen Sie das Handbuch.***

4.5.21 (DE) möchten / wollen / wünschen

Inhaltlich sind *wollen*, *möchten* und *wünschen* nahezu gleichbedeutend.

- *Möchten* ist die förmlichere und höflichere Form.
- *Wollen* drückt einen Wunsch stärker aus als *möchten*.
- *Wünschen* ist besonders emotional und daher in Technischer Dokumentation fehl am Platz.

✔ **Ja:** *Wenn Sie möchten, können Sie den Film auch rückwärts abspielen.*

(Vermutung: Der Wunsch hierzu ist eher schwach.)

✔ **Ja:** *Wenn Sie möchten, übernehmen wir alle Formalitäten für Sie.*

(Stärke des Wunsches unbekannt. Daher wurde die höflichere Form *möchten* bevorzugt.)

✔ **Ja:** *Klicken Sie nur dann auf die Schaltfläche Delete, wenn Sie die Datei tatsächlich löschen wollen.*

(Wer eine Datei löschen will, muss hierzu einen starken Wunsch haben.)

✘ **Nein:** *Wenn Sie eine Bestellbestätigung per E-Mail wünschen, aktivieren Sie das Kontrollkästchen Bestätigung versenden.*

✔ **Ja:** *Wenn Sie eine Bestellbestätigung per E-Mail erhalten möchten, aktivieren Sie das Kontrollkästchen **Bestätigung versenden**.*

5   References

This is the end of the book—but it's not the end of the story.

We hope that this book made you aware of the key factors that account for good user assistance, and we hope that the book will be your guide when you create your own documents.

If you want to learn more, take a look at the books in our Technical Documentation Solutions Series, which cover all aspects of creating user assistance in more detail.

- Technical Documentation Solutions Series: **Planning and Structuring User Assistance** — How to organize user manuals, online help systems, and other forms of user assistance in a user-friendly, easily accessible way

- Technical Documentation Solutions Series: **Designing Templates and Formatting Documents** — How to make user manuals and online help systems visually appealing and easy to read, and how to make templates efficient to use

- Technical Documentation Solutions Series: **Writing Plain Instructions** — How to write user manuals, online help, and other forms of user assistance that every user understands

- Technical Documentation Solutions Series: **Illustrating and Animating Help and Manuals** — How to create pictures, instruction videos, and screencasts that communicate technical information clearly

In addition, the following bibliography may be helpful. It also contains the books mentioned above. Flag symbols indicate the language of each book.

Books on technical documentation in general

Achtelig, Marc
Planning and Structuring User Assistance: How to organize user manuals, online help systems, and other forms of user assistance in a user-friendly, easily accessible way. indoition, 2012.

Achtelig, Marc
*Technical Documentation Essentials: "How to Write That F***ing Manual": The essentials of technical writing in a nutshell.* indoition, 2012.

Achtelig, Marc
*Technische Dokumentation: „How to Write That F***ing Manual": Ohne Umschweife zu benutzerfreundlichen Handbüchern und Hilfen.* Zweisprachige Ausgabe Englisch + Deutsch. indoition, 2012.

Achtelig, Marc
Translating Technical Documentation Without Losing Quality: What you shouldn't spoil when translating user manuals and online help. indoition, 2012.

Ament, Kurt
Indexing: A Nuts-and-Bolts Guide for Technical Writers. William Andrew, 2007.

Ament, Kurt
Single Sourcing: Building Modular Documentation. William Andrew, 2002.

Ballstaedt, Steffen-Peter
Wissensvermittlung. Die Gestaltung von Lernmaterial. Beltz Psychologische Verlags Union PVU, 1997.

Barker, Thomas T.
Writing Software Documentation: A Task-Oriented Approach. Longman, 2002.

Baumert, Andreas
Interviews in der Recherche: Redaktionelle Gespräche zur Informationsbeschaffung. VS Verlag für Sozialwissenschaften, 2004.

Bellamy, Laura; Carey, Michelle; Schlotfeldt, Jenifer
DITA Best Practices: A Roadmap for Writing, Editing, and Architecting in DITA. IBM, 2011.

Bellem, Birgit; Dreikorn, Johannes; Drewer, Petra; Fleury, Isabelle; Haldimann, Ralf; Jung, Martin; Keul, Udo P.; Klemm, Viktoria; Lobach, Sabine; Prusseit, Ines; Reuther, Ursula; Schmeling, Roland; Schmitz, Klaus-Dirk; Sütterlin, Volker
Leitlinie Regelbasiertes Schreiben – Deutsch für die Technische Kommunikation. tekom, 2010.

Bremer, Michael
The User Manual Manual: How to Research, Write, Test, Edit & Produce a Software Manual. Untechnical, 1999.

Brändle, Max (Hrsg.); Gabriel, Carl-Heinz (Hrsg.); Pforr, Reinhard (Hrsg); Pichler, Wolfram (Hrsg.); Schmidt, Curt (Hrsg.); Schulz, Matthias (Hrsg.)
Leitfaden für Betriebsanleitungen. tekom, 2010.

Carroll, John M. (Editor)
Minimalism Beyond the Nurnberg Funnel. The MIT Press, 1998.

Carroll, John M.
The Nurnberg Funnel: Designing Minimalist Instruction for Practical Computer Skill. The MIT Press, 1990.

Clements, Paul; Bachmann, Felix; Bass, Len; Garlan, David; Ivers, James; Little, Reed; Merson, Paulo; Nord, Robert; Stafford, Judith
Documenting Software Architectures: Views and Beyond. Addison-Wesley, 2010.

Clark, Ruth C.
Developing Technical Training: A Structured Approach for Developing Classroom and Computer-based Instructional Materials. Pfeiffer, 2007.

Clark, Ruth C.; Mayer, Richard E.
E-Learning and the Science of Instruction: Proven Guidelines for Consumers and Designers of Multimedia Learning. Pfeiffer, 2011.

Closs, Sissi
Single Source Publishing: Topicorientierte Strukturierung und DITA. entwickler press, 2006.

Coe, Marlana
Human Factors for Technical Communicators. Wiley, 1996.

Cowan, Charles
XML in Technical Communication. Institute of Scientific and Technical Communication, 2010.

DIN e.V (Herausgeber)
Technische Dokumentation: Normen für Produktdokumentation und Dokumentenmanagement. Beuth, 2008.

Drewer Petra; Ziegler, Wolfgang
Technische Dokumentation. Vogel Business Media, 2011.

Ferlein, Jörg; Hartge, Nicole
Technische Dokumentation für internationale Märkte: Haftungsrechtliche Grundlagen, Sprache, Gestaltung, Redaktion und Übersetzung. Expert, 2008.

Garrand, Timothy
Writing for Multimedia and the Web: A Practical Guide to Content Development for Interactive Media. Focal, 2006.

Gentle, Anne
Conversation and Community: The Social Web for Documentation. XML Press, 2009.

Grünwied, Gertrud
Software-Dokumentation: Grundlagen – Praxis – Lösungen. Expert, 2006.

Grupp, Josef
Handbuch Technische Dokumentation: Produktinformationen rechtskonform aufbereiten, wirtschaftlich erstellen, verständlich kommunizieren. Hanser, 2008.

Hackos, JoAnn T.
Information Development: Managing Your Documentation Projects, Portfolio, and People. Wiley, 2007.

Hackos, JoAnn T.
Introduction to DITA: A User Guide to the Darwin Information Typing Architecture. Comtech Services, 2011.

Hahn, Hans-Peter
Technische Dokumentation leichtgemacht. Hanser, 1996.

Hamilton, Richard L.
Managing Writers: A Real-World Guide to Managing Technical Documentation. XML Press, 2008.

Hargis, Gretchen; Carey, Michelle; Hernandez, Ann Kilty; Hughes, Polly; Longo, Deirdre; Rouiller, Shannon; Wilde, Elizabeth
Developing Quality Technical Information: A Handbook for Writers and Editors. IBM, 2004.

Hartman, Peter J.
Starting a Documentation Group: A Hands-On Guide. Clear Point Consultants, 1999.

Hennig, Jörg (Hrsg.), Tjarks-Sobhani, Marita (Hrsg.)
Arbeits- und Gestaltungsempfehlungen für Technische Dokumentation: Eine kritische Bestandsaufnahme. Schmidt-Römhild, 2008.

Hennig, Jörg (Hrsg.); Tjarks-Sobhani, Marita (Hrsg.)
Multimediale Technische Dokumentation. Schmidt-Römhild, 2010.

Hentrich, Johannes
DITA: Der neue Standard für Technische Dokumentation. XLcontent, 2008.

Hoffmann, Walter; Hölscher, Brigitte G.; Thiele, Ulrich
Handbuch für technische Autoren und Redakteure: Produktinformation und Dokumentation im Multimedia-Zeitalter. Publicis, 2002.

Hörmann, Hans
Meinen und Verstehen: Grundzüge einer psychologischen Semantik. Suhrkamp, 1978.

Horn, Robert E.
Mapping Hypertext: The Analysis, Organization, and Display of Knowledge for the Next Generation of On-Line Text and Graphics. Lexington, 1990.

Horton, William
Designing and Writing Online Documentation: Hypermedia for Self-Supporting Products. Wiley, 1994.

Johnston, Carol; Critcher, Ginny; Pratt, Ellis
How to write instructions. Cherryleaf, 2011.

Juhl, Dietrich
Technische Dokumentation: Anleitungen und Beispiele. Springer, 2005.

Kothes, Lars
Grundlagen der Technischen Dokumentation: Anleitungen verständlich und normgerecht erstellen. Springer, 2010.

Kühn, Cornelia
Handlungsorientierte Gestaltung von Bedienungsanleitungen. Schmidt-Römhild, 2004.

Muthig, Jürgen (Hrsg.)
Standardisierungsmethoden für die Technische Dokumentation. Schmidt-Römhild, 2008.

Pearsall, Thomas E.; Cook, Kelli Cargile
Elements of Technical Writing. Longman, 2009.

Price, Jonathan; Korman, Henry
How to Communicate Technical Information: A Handbook of Software and Hardware Documentation. Addison-Wesley Professional, 1993.

Pringle, Alan S.; O'Keefe, Sarah S.
Technical Writing 101: A Real-World Guide to Planning and Writing Technical Documentation. Scriptorium, 2009.

Rockley, Ann; Manning, Steve; Coopern Charles
Dita 101. lulu, 2009.

Rockley, Ann; Cooper, Charles
Managing Enterprise Content: A Unified Content Strategy. New Riders, 2012.

Schriver, Karen A.
Dynamics in Document Design: Creating Text for Readers. Wiley, 1996.

Schwarzman, Steven
Technical Writing Management: A Practical Guide. CreateSpace, 2011.

Self, Tony
The DITA Style Guide: Best Practices for Authors. Scriptorium, 2011.

tekom (Hrsg.)
Richtlinie zur Erstellung von Sicherheitshinweisen in Betriebsanleitungen. tekom, 2005.

Thiele, Ulrich
Technische Dokumentationen professionell erstellen. WEKA, 2009.

Thiemann, Petra; Krings, David
Creating User-Friendly Online Help: Basics and Implementation with MadCap Flare. CreateSpace, 2009.

Tuffley, Dr. David
Software User Documentation: A How To Guide for Project Staff. CreateSpace, 2011.

Van Laan, Krista; Julian, Catherine; Hackos, JoAnn
The Complete Idiot's Guide to Technical Writing. Alpha, 2001.

Weber, Jean Hollis
Is the Help Helpful? How to Create Online Help That Meets Your Users' Needs. Hentzenwerke, 2004.

Weber, Klaus H.
Dokumentation verfahrenstechnischer Anlagen: Praxishandbuch mit Checklisten und Beispielen. Springer, 2008

Weiß, Cornelia
Professionell dokumentieren. Beltz, 2000.

Weiss, Edmond H.
How To Write Usable User Documentation. Oryx, 1991.

Welinske, Joe
Developing User Assistance for Mobile Apps. Lulu, 2011.

Wieringa, Douglas; Barnes, Valerie E.; Moore, Christopher
Procedure Writing: Principles and Practices. Battelle, 1998.

Young, Indi
Mental Models: Aligning Design Strategy with Human Behavior. Rosenfeld, 2008.

Books on plain language and style

Achtelig, Marc
Writing Plain Instructions: How to write user manuals, online help, and other forms of user assistance that every user understands. indoition, 2012.

Achtelig, Marc
Writing Plain Instructions: Wie Sie Handbücher, Online-Hilfen und andere Formen Technischer Kommunikation schreiben, die jeder Benutzer versteht. Zweisprachige Ausgabe Englisch + Deutsch. indoition 2012.

Alred, Gerald J.; Brusaw, Charles T.; Oliu, Walter E.
Handbook of Technical Writing. St. Martin's, 2011.

Baumert, Andreas
Professionell texten: Grundlagen, Tipps und Techniken. dtv, 2008.

Blake, Gary; Bly, Robert W.
The Elements of Technical Writing. Longman, 2000.

Bremer, Michael
Untechnical Writing – How to Write About Technical Subjects and Products So Anyone Can Understand. UnTechnical, 1999.

Brogan, John A.
Clear Technical Writing. Career Education, 1973.

Burkhart, David
Stylistic traps in technical English – and solutions: Stilistische Fallen im Technischen Englisch – und Lösungen. BDÜ Fachverlag, 2010.

Bush, Donald W.
How to Edit Technical Documents. Oryx, 1995.

Goldstein, Norm
The Associated Press Stylebook and Briefing on Media Law. Basic Books, 2011.

Ikonomidis, Ageliki
Anglizismen auf gut Deutsch: Ein Leitfaden zur Verwendung von Anglizismen in deutschen Texten. Buske, 2009.

Jenkins, Jana; DeRespinis, Francis; Laird, Amy; Radzinski, Eric; McDonald, Leslie I.; Hayward, Peter
The IBM Style Guide: Conventions for Writers and Editors. Addison-Wesley Longman, 2011.

Kohl, John
The Global English Style Guide: Writing Clear, Translatable Documentation for a Global Market. SAS Press, 2008.

Langer, Inghard; Schulz von Thun, Friedemann; Tausch, Reinhard
Sich verständlich ausdrücken. Reinhardt, 2011.

Mackowiak, Klaus
Die 101 häufigsten Fehler im Deutschen: und wie man sie vermeidet. Beck, 2009.

Microsoft Corporation
Microsoft Manual of Style. Microsoft Press, 2011.

Rechenberg, Peter
Technisches Schreiben: (nicht nur) für Informatiker. Hanser, 2006.

Reiter, Markus; Sommer, Steffen
Perfekt schreiben. Hanser, 2009.

Ross-Larson, Bruce
Edit Yourself: A Manual for Everyone Who Works with Words. W. W. Norton, 1996.

Ross-Larson, Bruce
Writing for the Information Age. W. W. Norton, 2002.

Rothkegel, Annely
Technikkommunikation. UTB, 2010.

Schneider, Wolf
Deutsch für Profis: Wege zu gutem Stil. Goldmann, 2001.

Sick, Bastian
Der Dativ ist dem Genitiv sein Tod. Kiepenheuer & Witsch, 2004.

Strunk Jr., William
The Elements of Style. Longman, 1999.

Sun Technical Publications
Read Me First! A Style Guide for the Computer Industry. Prentice Hall, 2009.

University of Chicago Press
The Chicago Manual of Style. University of Chicago Press, 2010.

Weiss, Edmond H.
100 Writing Remedies: Practical Exercises for Technical Writing. Oryx, 1990.

Weiss, Edmond H.
The Elements of International English Style: A Guide to Writing Correspondence, Reports, Technical Documents, and Internet Pages for a Global Audience. M.E. Sharpe, 2005.

Weissgerber, Monika
Schreiben in technischen Berufen: Der Ratgeber für Ingenieure und Techniker: Berichte, Dokumentationen, Präsentationen, Fachartikel, Schulungsunterlagen. Publicis, 2010.

Books on graphics and design

Achtelig, Marc
Designing Templates and Formatting Documents: How to make user manuals and online help systems visually appealing and easy to read, and how to make templates efficient to use. indoition, 2012.

Alexander, Kerstin
Kompendium der visuellen Information und Kommunikation. Springer, 2007.

Ballstaedt, Steffen-Peter
Visualisieren: Über den richtigen Einsatz von Bildern. UTB, 2011.

Clark, Ruth C.; Lyons, Chopeta
Graphics for Learning: Proven Guidelines for Planning, Designing, and Evaluating Visuals in Training Materials. Pfeiffer, 2010.

Cooper, Alan
About Face: The Essentials of User Interface Design. IDG, 1999.

Cooper, Alan
The Inmates Are Running the Asylum. SAMS, 1999.

Hennig, Jörg (Herausgeber); Marita Tjarks-Sobhani (Herausgeber)
Visualisierung in Technischer Dokumentation. Schmidt-Römhild, 2004.

Horton, William
Illustrating Computer Documentation: The Art of Presenting Information Graphically on Paper and Online. Wiley, 1991.

Runk, Claudia
Grundkurs Grafik und Gestaltung. Galileo, 2010.

Williams, Robin
The Non-Designer's Type Book. Peachpit, 2005.

Williams, Robin
The Non-Designer's Design Book. Peachpit, 2008.

Wirth, Thomas
Missing Links: Über gutes Webdesign. Hanser, 2004.

6 Feedback

We sincerely hope that reading this book was a rewarding experience.

- If you like this book and think that it can help you improve your own documents, please don't hesitate to post a review and recommend the book to your colleagues. Also, don't hesitate to drop us a line. It motivates us so much to carry on :-).

- If you didn't like this book—we're embarrassed and awfully sorry. Could you please send us some feedback about what you think we should improve?

Our email address is: *feedback-ESS-DE-1@indoition.com*

If you'd like to help us even more, please also email us your answers to the questions below.

Thank you for your support.

How to answer the questions

Please email your answers to:
feedback-ESS-DE-1@indoition.com

For example, your email could look like this:
1c, 2a, 3b, 4a, 5d, 6a, 7a, 8b, 9b, 10a, 11a, 12b

We won't use your data for any purpose other than improving future editions of this book. If you don't want to answer all questions, that's perfectly OK. Just answer the ones that you feel comfortable with.

1. Questions about the book

How did you feel about the length of the book?

It was much too long.	1a
It was slightly too long.	1b
It was just perfect.	1c
It was slightly too short.	1d
It was much too short.	1e

Did the book cover what you'd expected, based on its title and description?

I didn't miss anything.	2a
I missed a few minor things.	2b
I missed some important points.	2c

How did you experience the depth of information?

Much of the presented information was too trivial for me.	3a
The information was just what I needed.	3b
Much of the information was too specialized for me.	3c

How did you like the practical nature of the book?

I appreciated the lack of theory and technical terms.	4a
I missed scientific background information, references to studies, and more precise terminology.	4b

Did you find any mistakes?

Yes, too many.	5a
Some, but not more than usual.	5b
Only very few.	5c
None.	5d

(If your answer is "None": Go through the book again before answering this question! No book is free from errors. If you have the time, please tell us more about the mistakes that you've found.)

2. Questions about your professional background

What's your main professional occupation?

technical writing	6a
support	6b
development	6c
marketing	6d
product management	6e
translation	6f
other	6g

How many years of experience in technical writing do you have?

less than 1	7a
1 to 3	7b
more than 3	7c

Which kind of products do you document?

mainly hardware	8a
mainly software	8b
a mixture of both hardware and software	8c

Who reads the documents that you write?

mainly consumers	9a
mainly professional users	9b

Do you speak English as a first language?

English is my first language.	10a
I speak English as a second language.	10b

Do you mainly write in English?

Yes, more than 50% of my texts are in English.	11a
No, less than 50% of my texts are in English.	11b
No, I don't write English documents at all.	11c

Who purchased this book?

I purchased the book at my own expense.	12a
The organization that I work for purchased the book.	12b
I borrowed the book from a colleague.	12c
I borrowed the book from a public library.	12d
I received a copy in a training course.	12e

Index

S

Weitere Bücher von indoition publishing zum Thema Technische Dokumentation:

„Writing Plain Instructions"

Wie Sie Handbücher, Online-Hilfen und andere Formen Technischer Kommunikation schreiben, die jeder Benutzer versteht

Zweisprachige Ausgabe: Englisch + Deutsch

„Planning and Structuring User Assistance"

Wie Sie Handbücher, Online-Hilfen und andere Formen Technischer Dokumentation benutzerfreundlich aufbauen und den Informationszugriff erleichtern

Zweisprachige Ausgabe: Englisch + Deutsch

„Designing Templates and Formatting Documents"

Wie Sie Benutzerhandbücher und Online-Hilfen attraktiv und gut lesbar gestalten, und wie Sie effiziente Formatvorlagen erstellen

Zweisprachige Ausgabe: Englisch + Deutsch

„Illustrating and Animating Help and Manuals"

Wie Sie Bilder, Instruktionsvideos und Screencasts erstellen, die technische Informationen verständlich vermitteln

Zweisprachige Ausgabe: Englisch + Deutsch

„Dokumentation verlustfrei übersetzen"

Was Sie beim Übersetzen von Benutzerhandbüchern und Online-Hilfen nicht zerstören sollten

Zweisprachige Ausgabe: Englisch + Deutsch

Detaillierte Informationen zu allen Ausgaben finden Sie unter *www.indoition.de*.

Copy and Paste Kit Technische Dokumentation
Ihre Bausteine für verständliche Dokumentation

Das Technical Documentation Copy and Paste Kit ist eine wesentlich erweiterte Online-Version dieses Buchs sowie aller Bücher der Reihe „Lösungen zur Technischen Dokumentation". Das Kit ist Ihr steter Begleiter und Styleguide (Redaktionsleitfaden) während **aller Phasen eines Dokumentations-Projekts**:

- Analyse der Anforderungen
- Strukturierung der Inhalte
- Design von Formatvorlagen
- Schreiben der Texte
- Anfertigen von Bildern
- Lektorat und Korrektur
- Übersetzung

Wie dieses Buch, enthält auch das Kit keine langen theoretischen Abhandlungen. Dafür bietet es praxisnahe Empfehlungen und Beispiele, die Sie einfach kopieren und direkt auf Ihre eigene Arbeit anwenden können.

Sie können das Kit entweder lokal, auf einem Netzlaufwerk oder auf einem Webserver installieren. Somit haben Sie und Ihr Team **jederzeit und von überall aus darauf Zugriff**.

Sie können sogar **Ihre eigenen, firmenspezifischen Anmerkungen und Spezifikationen hinzufügen**. Um Ihre Anmerkungen zu bearbeiten und zu verwalten, können Sie nahezu jeden HTML-Editor, ein Wiki, ein Commenting Script oder ein Content-Management-System verwenden. Sie können entweder allen Mitgliedern Ihres Teams das Bearbeiten von Notizen und Kommentaren erlauben oder hierfür einen Moderator bestimmen. Wenn Sie ein Update des Kits installieren, bleiben Ihre Notizen und Kommentare vollständig erhalten.

Auf diese Weise erhalten Sie de facto Ihren eigenen, **firmenspezifischen Styleguide**, ohne sich in wesentlichen Teilen um dessen Erstellung und Pflege kümmern zu müssen.

Mehr Informationen sowie eine Demo finden Sie unter *www.indoition.de*.

indoition Hotkey Script Collection for Writers and Translators

Zeitsparende Makros zum Schreiben und Nachschlagen in jedem Programm

Die Scripts der „indoition Hotkey Script Collection for Writers and Translators" machen Ihre Arbeit effizienter:

- Geben Sie **häufig verwendete Wörter und Wendungen automatisiert** mit Hilfe bestimmter Tastenkombinationen ein.

- **Tippen Sie Sonderzeichen komfortabel mit einem einzigen Tastendruck**, z. B.: sprachspezifische Sonderzeichen, diakritische Zeichen, typografische Anführungszeichen, typografische Apostrophe und Gedankenstriche.

- **Verwandeln Sie die Feststelltaste** in eine reguläre Umschalttaste, so dass SO ETWAS NIE MEHR PASSIERT, wenn Sie versehentlich die Feststelltaste drücken.

- **Schlagen Sie ein markiertes Wort per Tastendruck** in jedem beliebigen Online-Wörterbuch oder Online-Lexikon **nach**.

- Und vieles mehr ...

Alle Scripts können Sie bei Bedarf einfach bearbeiten und individuell anpassen. Dazu brauchen Sie keine fortgeschrittenen Programmierkenntnisse.

Anders als Makros, die für eine bestimmte Applikation programmiert wurden (wie z. B. Microsoft-Word-Makros), funktionieren die Scripts der Script Collection in *allen* Programmen für Windows.

Mehr Informationen sowie eine Demo finden Sie unter *www.indoition.de*.

indoition Starter Template

Professionelle Formatvorlage für Technische Dokumentation

Viele Autorenwerkzeuge werden ohne geeignete Formatvorlagen zum Erstellen klarer, ansprechender Technischer Dokumentation geliefert. Eine eigene Formatvorlage komplett neu zu erstellen, kann zeitaufwendig sein. Das „indoition Starter Template" beschleunigt diese Arbeit und hilft Ihnen, kostspielige strategische Fehler von vornherein zu vermeiden. Es bietet:

- Ein Design, das nicht nur angenehm fürs Auge ist, sondern auch Ihre Inhalte klar kommuniziert.
- Absatzformate und Zeichenformate, die beim Schreiben Ihrer Dokumente einfach anzuwenden sind.

Das Starter Template wurde für Microsoft Word, OpenOffice und LibreOffice und die Papierformate A4 und Letter entwickelt. Falls Sie ein anderes Papierformat nutzen, müssen Sie im Wesentlichen nur die Einstellungen für die Seitenränder ändern.

Auch viele andere Autorenwerkzeuge können Microsoft-Word-Dateien (*.docx) und die von OpenOffice / LibreOffice genutzten OpenDocument-Text-Dateien (*.odt) importieren.

Wesentliche Merkmale:

- **Kein Schnickschnack** – die Formatvorlage enthält nur das, was Sie und Ihre Leser wirklich brauchen.
- **Automatisierte Formate** machen manuelle Formatierungen weitgehend überflüssig. Optimierte Einstellungen sorgen für automatische Zeilenumbrüche und Seitenumbrüche.
- **Bewährtes**, systematisches Schema für Formatnamen und Tastenkürzel.
- **Funktioniert mit allen Sprachversionen von Microsoft Word, OpenOffice und LibreOffice**. Sie müssen keine Formatnamen und Feldfunktionen anpassen.
- **Vorbereitet** auf die Möglichkeit, aus Ihren Dokumenten mit Hilfe eins geeigneten Single-Source-Publishing-Tools oder Konverters auch **Online-Hilfen** erzeugen zu können.
- Enthält eine **ausführliche Anleitung**, wie Sie die Formate anwenden und bei Bedarf ändern können.

Mehr Informationen unter *www.indoition.de*.

This is *not* a happy customer:

Make them happy, write better help!

Help+Manual®

www.helpandmanual.com

Help+Manual creates all
standard online help formats
including **HTML Help, Webhelp,
PDF manuals** and **e-books**
from one single source.

And it's as easy to use as
a word processor.

Learn more on our website
http://www.helpandmanual.com!

www.ingramcontent.com/pod-product-compliance
Lightning Source LLC
La Vergne TN
LVHW022305060326
832902LV00020B/3293